BAND THEORY OF METALS

BAND THEORY OF METALS

THE ELEMENTS

BY

SIMON L. ALTMANN

*Lecturer in the Theory of Metals and
Fellow of Brasenose College, University of Oxford*

PERGAMON PRESS

Oxford · New York · Toronto · Sydney · Braunschweig

Pergamon Press Ltd., Headington Hill Hall, Oxford
Pergamon Press Inc., Maxwell House, Fairview Park, Elmsford, New York 10523
Pergamon of Canada Ltd., 207 Queen's Quay West, Toronto 1
Pergamon Press (Aust.) Pty. Ltd., 19a Boundary Street,
Rushcutters Bay, N.S.W. 2011, Australia
Vieweg & Sohn GmbH, Burgplatz 1, Braunschweig

First edition 1970
Library of Congress Catalog Card No. 70−112612

Printed in Great Britain by Willmer Brothers Limited, Birkenhead

08 015602 9 (hard cover)
08 015601 0 (flexicover)

Contents

3. The Effect of the Crystal Field in One Dimension: Bloch Functions   62

Preface

THE topics discussed in this book cover on purpose a very limited part of the band theory of solids. Although solid state physics is admirably treated in a number of books, beginners often find it difficult to grasp its formal principles, in particular the theory and properties of Brillouin zones. So I have endeavoured to provide a thoroughly self-contained account of these principles in a form suitable for undergraduates with a limited experience of mathematics and quantum mechanics, such as metallurgists and experimental physicists. A little knowledge of crystallography is assumed, in particular some familiarity with the body centred cubic, face centred cubic, and hexagonal close-packed structures. On the other hand, the reciprocal lattice is fully discussed.

Very few applications are treated in this book since, after reading it, the student should have no difficulty in following the various treatments of them already available. Of course, some applications have been included in order to illustrate the use of the formal ideas developed. Thus a detailed account is given of the computation of the Fermi surface of copper and its comparison with experiment.

The book has been programmed for self-study and has thus been made as self-contained as possible: an account is given, for instance, of the major ideas and methods of quantum mechanics that are used. All steps in the proofs are given in detail and the subjects have been tackled in an order designed so as to build up skill and self-confidence in the reader before the more difficult topics are treated. Thus a thorough study is made of one-dimensional systems before three-dimensional ones are discussed. Markings are provided to guide the reading and a number of sections have been included for those who may wish to acquire some manipulative experience in the subject. However, the body of the book has been made entirely independent of these sections, which are marked with "bullets" (●).

ix

On account of the level desired for the book, the use of group theory was ruled out. Nevertheless, the concept of symmetry has been taken as the backbone of the book, and I hope that this form of approach may prove useful for readers who will pursue further the study of the theory of solids. Indeed, some subjects are given in the bulleted sections that are not normally treated in elementary textbooks. For instance, the sticking together of bands on the hexagonal faces of the hexagonal close-packed Brillouin zone is fully discussed from the properties of the complex conjugation operator.

Here and there I have not been as rigorous as I would have wished. In one or two cases I had not developed the mathematical apparatus required, as in § **3.1.3**, where the commutation of symmetry operators with the hamiltonian would normally be derived from the invariance of the hamiltonian matrix under the unitary transformation effected. In others, the subtleties of an accurate treatment would not have made sense to the reader in the context. Thus the transformation of functions under a symmetry operation in § **3.1.2** is presented naïvely, whereas this definition as well as the distinction between the active and passive pictures of symmetry operations (which is not made in the book) require delicate discussion in group theory. I confess unblushingly to these shortcomings: not only the rigour achieved would have been unrelated to the level of the work done in the book, but it would have distracted the reader from the main argument.

It is not without trepidation that I face the fact that this must be the only book in solid state where a plane wave is written $\exp(2\pi ikx)$ instead of $\exp(ikx)$. It is far from my intention to persuade solid state practitioners to change their notation. It must be recognized, however, that the saving of 2π in the plane waves requires its grafting into the reciprocal space, so that the 2π's appear at least as often and where they hurt more, since the reciprocal lattice is a notorious stumbling block for beginners. Even more seriously, the definition of the reciprocal lattice becomes different from the standard one in crystallography, with the result that mistakes are sometimes made. I hope that the use of the standard crystallographic reciprocal lattice throughout this book (with careful indication of the changes required in order to transfer to the solid state notation) will make the reader aware of the points where care must be exercised.

This text grew out of a course of eight to twelve lectures, covering essentially the unbulleted sections of this book, which I have given at Oxford and elsewhere. In writing it up, I have added a certain amount of material that answers the many questions asked by my students after the lectures. This has made the discussion here and there less light than I would have wished, but I gave priority to the need to remove loopholes from the treatment. In lectures, of course, I would still prefer to gloss over some of the more subtle points in order to stress the more important features.

This book was started owing to the prompting of Professor W. Hume-Rothery who, in fact, asked many of the questions which I have tried to answer. It is my sorrow that he did not live to see the book published. I should like to thank many friends also for their help, in particular Dr. Christopher Bradley, who read the manuscript and eliminated some inaccuracies. A first draft of the book was written for a course of lectures that I delivered at the University of Rome at the kind invitation of Professors V. Caglioti and G. Sartori.

Oxford

How to Use this Book

CROSS-REFERENCES AND FORMULAE

In cross-references to other chapters, the number of the chapter is the first numeral given in bold type. Thus, § **3**.2 means § 2 of Chapter 3 and (**3**.2) means equation 2 of Chapter 3. References within a chapter carry no chapter numeral. Thus, § 2 and (2) mean respectively § 2 and equation (2) of the current chapter.

HEADING MARKINGS

Markings such as ($\rightarrow$ **3**) or (**2**.18 $\leftarrow$) after a heading mean respectively that the contents of the section in question will not be required until Chapter 3 or that it is required only to provide a proof of a result already used in § 18 of Chapter 2. Such sections can be skipped in a first reading, a procedure which is generally recommended.

BULLETS (●)

A number of sections of the book are interconnected amongst themselves and so arranged that the rest of the book is independent of them. They are marked with bullets so that all sections having the same number of bullets belong to a given interconnected set of sections. In general, bulleted sections are designed so as to provide some manipulative skill at handling the various concepts which are discussed more qualitatively in the rest of the book. Readers who are only interested in the basic concepts can leave out these sections altogether. Others may wish to leave them for a second reading, a procedure which is recommended.

EXERCISES AND PROBLEMS

Exercises are pieces of detailed analysis fairly fully worked out. They are separated as such to permit a terser form of presentation and to indicate that the reader may have to do some of the work. Problems are left to the reader, but answers and hints are provided if needed.

PROOFS

From time to time a statement is first made and then proved or discussed. In order to warn the reader that this is so and that the statement in question is neither a consequence of a previous one nor meant to be merely accepted, such statements are immediately followed by sentences beginning with the words "in fact".

NOTATION

Vectors are represented in **bold face** and operators in **sanserif**, thus **R** and **R** respectively. R and $|\mathbf{R}|$ are both the modulus of **R**, the second symbol being used when greater emphasis is required. **m** is the electron mass and **e** the electron charge.

WARNING. ASTERISKS

This book uses one piece of notation which differs from that adopted in most books and papers on the subject.

A plane wave [see (**1.15**)] is written as $\exp(2\pi ikx)$ where $k = 1/\lambda$ is the wave number. The common usage in solid state is to write it as $\exp(ikx)$ where $k = 2\pi/\lambda$ is the *circular wave number*. As a result, the *k space* used in this book (see § 3.10) differs from the commonly used k space which, for convenience, will be referred to as *circular k space*. All lengths in circular k space are 2π times lengths in k space. The most important formulae or expressions for which changes are required are marked with an asterisk either on the equation number or at the end of the corresponding line. Unless a factor is explicitly given, the asterisk means that the right-hand side of the expression so marked must be multiplied by 2π in order to obtain the expressions commonly used in circular k space.

For the purpose of this warning the words "k space" and "reciprocal space" can be interchanged.

CHAPTER 1

Revision of Quantum Mechanics

Iт is not the purpose of this chapter to give an introduction to quantum mechanics: the reader will be expected to have some familiarity with the qualitative features of such concepts as wave functions and wave equations, as usually provided by elementary courses on atomic theory. Instead, a few points will be discussed that will have a central importance in the arguments to be used later on in this book.

In order to do this, the development of the operational ideas of quantum mechanics will be sketched from a few basic experimental facts (§ 1) and a fundamental principle (§ 2). It will thus be seen that the abstract concepts that are introduced are no more than a form of description of experimental facts, just as in any other physical theory, and that the postulates required have a direct physical content: any mysticism with which the reader may have endowed his views of the subject is not intrinsic to the theory.

1. Basic experimental facts

We shall accept the following as proven experimental facts.

(i) The existence of particles associated with a wave. Such a concept was first postulated by Einstein (1905) who introduced the *photons* in order to interpret the photoelectric effect. The experiments indicate that the energy E of the photons is given by

$$E = h\nu, \tag{1}$$

where $h = 6 \cdot 62 \times 10^{-27}$ erg sec (Planck's constant) and ν is the frequency of the wave. (1) is called *Planck's relation*.

(ii) The existence of a wave associated with a beam of particles. This was first demonstrated experimentally by Davisson and Germer

1

(1927) who observed that a beam of electrons of momentum p ($p = mv$, where m = mass, v = velocity) was diffracted when reflected by a metal surface, the periodic structure of which acts as a grating, and that the interference lines observed corresponded to those produced by a wave of wave number k ($k = \lambda^{-1}$, where λ = wavelength) which the experiment showed to be related to p as follows:

$$p = hk. \tag{2}$$

This is called the *de Broglie relation*.

2. Heisenberg's uncertainty principle

The two facts given above established the apparently paradoxical *wave particle duality*: an electron must be described both as a wave (wave number k) and as a particle (of momentum p). This paradox was resolved by Heisenberg who showed that a classical–mechanical description of the electron is not possible. Hence, although the electron

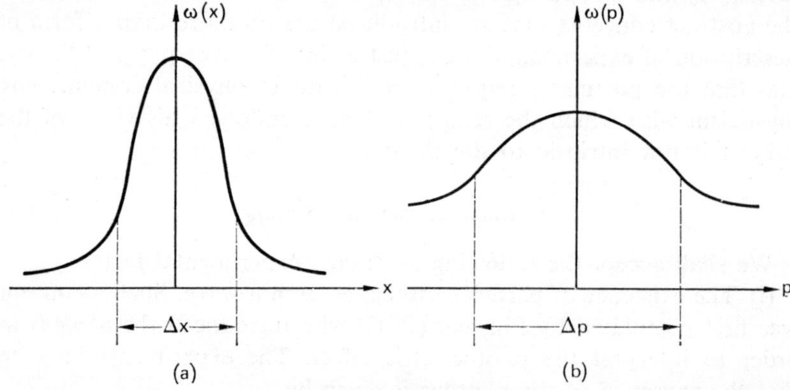

Fig. 1. Probability distributions.

may retain some of the attributes of a classical particle (such as the possession of a momentum), it can be expected to present new properties (such as a wavelike behaviour).

A classical particle is described by specifying precise simultaneous values of its momentum p and its position x: this, Heisenberg showed, can no longer be the case for an electron. In fact, suppose that we want to measure the position of an electron very precisely: we must illuminate it with light of very short wavelength, since it is well known from optics that the shortest length determinable with a wave of wavelength λ is λ. Hence $\nu = c\lambda^{-1}$ (c = velocity of light) must be large. From (1) it follows that the energy of the photons present in the light beam must be large and hence, when they impinge on the electron, they must change substantially its momentum. Let Δp and Δx be the error in the measurements of p and x respectively. The smaller Δx is, the larger Δp must be, since a small Δx requires a small λ, that is a photon of large energy that will introduce a large change Δp in the electron momentum. The relation between Δp and Δx must therefore be of the form $\Delta p \, \Delta x$ = a finite constant.

That is, if we know the position of the electron precisely ($\Delta x = 0$) its momentum is indeterminate ($\Delta p = \infty$). By means of a more detailed argument Heisenberg obtained the expression

$$\Delta p \, \Delta x = h, \tag{3}$$

which is called *Heisenberg's relation*. Heisenberg's *uncertainty principle* states that, given a value Δx, it is not possible to measure p with an error Δp smaller than that given by (3), and vice versa for x. This principle which, of course, is not directly verifiable experimentally, occupies the same position in quantum mechanics as the second principle does in thermodynamics (impossibility of construction of a perpetuum mobile of second class).

A more precise way of looking at relation (3) is as follows. Since each measurement of a variable such as x involves an unavoidable perturbation of the system measured, the results of a large number of measurements must be spread, i.e. different results x_i will appear, each of them repeated a different number of times n_i. We shall call $\omega(x_i) = n_i/n$ (n = total number of measurements) the *frequency* or *probability* of the value x_i. When we plot $\omega(x)$ as a function of x we expect to get the usual gaussian *probability distribution* curve of Fig. 1a, where, as in the theory of errors, the error Δx is measured by the distance between the turning points of the gaussian. Similarly, a probability distribution for p will be obtained, such as that represented in Fig. 1b. It follows from Heisenberg's principle that these two curves are not independent. Rather, Δp and Δx are related by (3), which requires that, if one of the distributions is very narrow, the other one must be very flat.

3. The state function

We define in classical mechanics the state Φ of a system as some function of the values of p and x for all the particles of a system: $\Phi = \Phi(p, x)$. This is so because we know from Newton's equations that, given p and x at the time t, their values at a later time t' are determined. Hence, given Φ_t (the value of Φ at the time t), we can always calculate $\Phi_{t'}$.† In practice, the definition of a *state function* is always arrived at by a process of trial and error: one has to find the smallest number of variables which describe a system and that are self-predicting, that is such that equations can be found which permit the calculation of the values of all the chosen variables at a time t' from the known values of the same variables at the time t. These variables are called *state variables* and a function Φ of them which appears in the equations just mentioned is a state function, the basic property of which is that Φ_t determines $\Phi_{t'}$, a relation which we shall denote symbolically as follows: $\Phi_t \rightarrow \Phi_{t'}$.

In the rest of this section we shall be concerned with the definition of the state function in quantum mechanics, which we shall denote with the symbol ψ, a more detailed specification of which will come later: it will be enough for the time being to state that ψ_t must be such that $\psi_t \rightarrow \psi_{t'}$.

We now start the trial and error process to determine ψ: in order to keep to a simple experimental situation we shall try to describe a beam of particles such as the one used in the Davisson and Germer experiment. We know that we shall not be able to predict any more than probability distributions of p and x. Suppose first that we try to identify ψ_t with $[\omega(x)]_t$ (Fig. 2). This will not do: in the Davisson

---

† The reader may not have used before this type of abstract presentation and may be baffled by the meaning of Φ. This, however, should not be difficult to visualize. The kinetic energy of a particle $\frac{1}{2}mv^2$ can be written as $p^2/2m$ and its potential energy is in general some function of position $V(x)$. Its total energy $E = p^2/2m + V(x) = E(p, x)$ is a state function in the sense described in the text and $\Phi(p, x)$ can be taken to be just this function: whenever the reader has used the principle of energy he has used a state function. (See also § 9.)

and Germer experiment we know without any doubt the direction in which the particles are moving, which we shall assume to be from left to right along the x axis of the figure, whereas there is nothing in $[\omega(x)]_t$ to tell us anything about that direction. To remedy this deficiency we might try to incorporate $[\omega(p)]_t$ as well as $[\omega(x)]_t$ in the definition of ψ_t, so that the first of these two distributions will provide information about the momentum and hence about the direction of

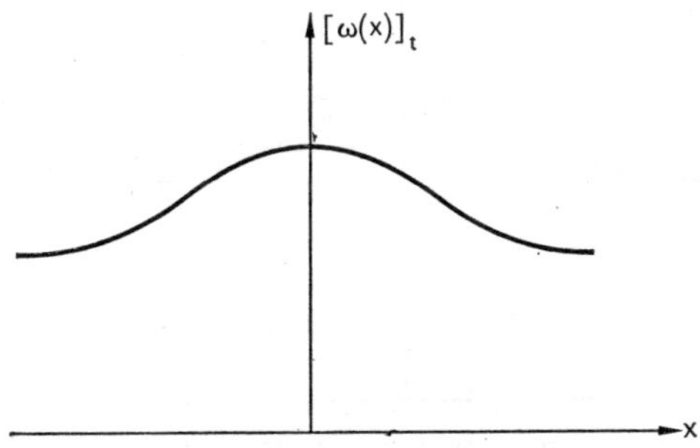

FIG. 2. A probability distribution at the
time t.

motion of the particles. (Remember that the momentum, like the velocity, is a vector: a positive momentum means that the particles move towards the right in Fig. 2.) This remedy, however, cannot be the right one because $\omega(p)$ and $\omega(x)$, on account of Heisenberg's principle, are not independent: it is an important requirement that all pieces of information about a system that are built into the state function be independent. (Redundant information is otherwise used, which is equivalent to not having a minimal set of variables as was required of the state variables.)

We next try to build up ψ_t from the two values of $\omega(x)$ at the two times t and $t+\delta t$ ($\delta t \rightarrow 0$), such as are represented in Fig. 3. In fact, the figure shows at once that the particles are moving to the right, as we know they are. It can readily be noticed that to know ω_t and $\omega_{t+\delta t}$ is equivalent to knowing ω_t and $[d\omega/dt]_t \equiv \dot{\omega}_t$ (the derivative of ω with respect to t at the time t). Hence, we can provisionally agree that ψ_t will be determined by $[\omega(x)]_t$ and $[\dot{\omega}(x)]_t$.

If this is so, the state function will depend on two functions and we must be very careful about our book-keeping: ω and $\dot\omega$ must somehow be kept separate enough so that ω alone can be recovered from ψ when required. (We have said that we must be able to predict probability distributions such as ω and if we were able to effect the prediction $\psi_t \to \psi_{t'}$, we must still be able to obtain $\omega_{t'}$ from $\psi_{t'}$.)

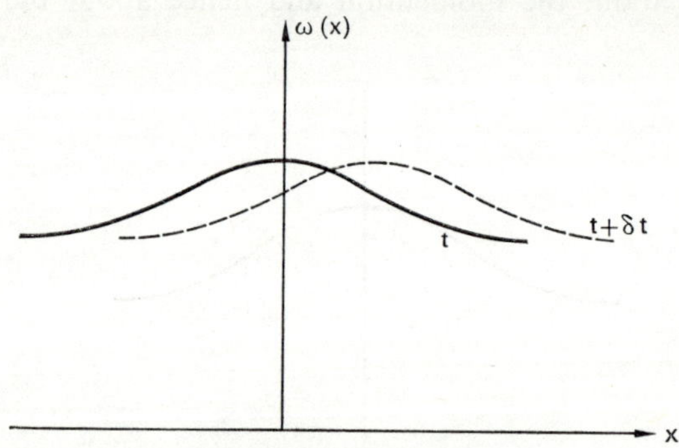

Fig. 3. Probability distributions at the
time t and $t+\delta t$.

It often happens in physics that two functions must be kept in a special relation of this type, and the simplest way to do this is always to build them up into a complex function. (In principle, for instance, ω and $\dot\omega$ could be, respectively, the real and imaginary parts of such a function, so that they could be easily identified at all times. In practice, it turns out that the way in which they are built up into ψ is more complicated: ω, as we shall see later, is the square of the modulus of ψ, whereas $\dot\omega$ is related to its argument.)

We shall now summarize the conditions to be satisfied by ψ:

Errata

(i) ψ is complex;   (iii) $\psi_t \to \psi_{t'}$;

(ii) $\psi_t \to \omega_t$;   (iv) $\omega_t \not\to \psi_t$.

(We use the arrow here in a wider sense than before, to mean "determines" and a crossed arrow to mean "does not determine".) The first three conditions follow from our previous discussion, and we justify (iv) as follows. If the state function could be determined from the probability distribution (i.e. if $\omega_t \to \psi_t$), we could build up the following chain:

$$\text{(a)} \quad \text{(b)} \quad \text{(c)}$$

$$\omega_t \to \psi_t \to \psi_{t'} \to \omega_{t'},$$

which entails $\omega_t \to \omega_{t'}$. If this were so, however, ω itself would be the state function (remember that its main property is to be self-predicting), whereas we noticed before that this identification is impossible. It follows that at least one link of the chain above must break: however (b) and (c) are correct [being conditions (iii) and (ii) respectively], hence process (a) must be impossible.

We can now continue with the trial and error process involved in finding the state function. Essentially, we must guess a form of $\psi(x)$ such that (i), (ii) and (iv) are satisfied and we must try to find an equation to carry out process (iii). The latter is, of course, the hard step: it is analogous to finding Newton's equations of motion in classical mechanics. However, since the corresponding equation (which is called the *Schrödinger equation with time*) is not one that we shall require much in our later work, we shall not concern ourselves with its derivation.

Returning to the guessing of $\psi(x)$, we shall now see that the following trial definition of it satisfies (i), (ii), and (iv):

$$\psi^*(x)\,\psi(x) = \omega(x), \tag{4}$$

where $\psi^*(x)$ is the complex conjugate of $\psi(x)$. (Notice that if $z = a + ib$, $z^*z = a^2 + b^2$, the square of the modulus of z, is always real, so that the right-hand side of (4) is real, as it should be.) That (i) and (ii) are satisfied follows immediately. (iv) is also satisfied.

Proof. Suppose that, for a given $\omega(x)$, we find a function $\psi(x)$ that verifies (4). Any function φ of the form $\varphi(x) = \exp(i\alpha)\psi(x)$, for any arbitrary value of α also verifies (4). In fact,

$$\varphi^*(x)\varphi(x) = \exp(-i\alpha)\psi^*(x)\,\exp(i\alpha)\psi(x) = \psi^*(x)\psi(x) = \omega(x).$$

Hence, a given $\omega(x)$ cannot determine $\psi(x)$ uniquely, as required by (iv).

In fact (4) is the simplest definition that satisfies conditions (i), (ii), and (iv). We shall accept it as the definition of the state function without proving that the vital condition (iii) is also satisfied. For reasons that will become clear in § 7 the quantum mechanical state function is usually called the *wave function*.

4. Stationary states

When the probability distribution $\omega(x)$ is independent of the time we say that the state is *stationary*.

All elementary applications of quantum mechanics in chemistry and solid state theory concern stationary states. This is not an unmixed blessing since, if only stationary states are considered, it is possible to fall into erroneous ideas about the nature of the wave function. This, we said, is determined at a given time by the values of $\omega(x)$ and $\dot{\omega}(x)$: the latter vanishes for a stationary state and hence there is no essential distinction in such a case between $\psi(x)$ and $\omega(x)$. Trivially, $\omega_t \rightarrow \omega_{t'}$, since both are equal. It is important for the reader to realize that these properties are valid only for stationary states, and that since the state function is essentially a predictive instrument its true flavour cannot be fully appreciated until one handles time dependent problems. Nevertheless, stationary states have some remarkable properties in quantum mechanics, with which we shall be particularly concerned.

5. Normalization

Equation (4) can be interpreted as follows: the square of the modulus of the state function at a point x gives the probability of finding the particle at that point. We must now polish up this definition a little. The main difficulty is that the probability of finding the particle *exactly* at a point x is nil, since a mathematical point has no extension. Because of this, when dealing with continuous variables, probabilities are always replaced by *probability densities*, which is done as follows. If $n(x)$ is the number of measurements which yield results that appear in a small interval Δx around x, and n is the total number of measurements, the *probability density* $\omega(x)$ is the frequency of these results per *unit interval*:

$$\omega(x) = \frac{1}{\Delta x}\frac{n(x)}{n}.$$

More rigorously, we must take the limit, in the right-hand side of this expression, for $\Delta x \to 0$. It follows from this definition that the probability, $n(x)/n$, of finding the particle in the small interval dx (which we conveniently write as the limit of $\Delta x \to 0$) is $\omega(x)\,dx$.

We must now agree that $\omega(x)$ in (4) must be understood as a probability density.

Since the probability of finding the particle in either of the two intervals dx_i and dx_j is the sum of the respective probabilities, that of finding the particle somewhere on the x axis must be the sum of the probabilities $\omega(x)\,dx$ for all x, which of course will be an integral:

$$\int_{-\infty}^{\infty} \omega(x)\,dx = \int_{-\infty}^{\infty} \psi^*(x)\,\psi(x)\,dx \equiv N. \tag{5}$$

In principle, we ought to have $N = 1$, since it is certain that the particle will be found somewhere on the x axis (remember that, for simplicity, we are so far dealing with one-dimensional problems). In practice, this is not essential. If $N = 1$ we say that the state function is *normalized*. If $N \neq 1$, we can easily redefine (4) as follows:

$$\omega(x) = \frac{\psi^*(x)\,\psi(x)}{\displaystyle\int \psi^*(x)\,\psi(x)\,dx}, \tag{6}$$

and it is quite clear that $\int \omega(x)\,dx = 1$, as it should be. The denominator in (6) is called the *normalization integral*.

An important result should be clear from (6), namely that a state function can be multiplied by any arbitrary constant (c, say), without affecting at all its physical meaning, since both ψ and $c\psi$ correspond to the same value of the probability density $\omega(x)$.

6. Operators and eigenvalue equations

Suppose that we have a system with state function ψ and that we measure a variable a (such as energy, momentum, etc.) in the system. Since the measurement perturbs the system, the state function must be changed from ψ into ψ'. However, if we immediately repeat the measurement of the same variable, ψ' is not perturbed and we must obtain the same result for the variable a (just as in classical physics). In fact, an example of this rule is illustrated in Fig. 4, for an electron beam with a large dispersion of the position coordinate on the xy plane perpendicular to the direction of the beam. The first measurement of the x,y coordinates of the electron (by means of the first diaphragm) perturbs the system, which is nevertheless unaltered by the second measurement.†

If we now use an arrow to mean "is transformed into", the measurement process just described can be represented as follows:

$$\text{meas. of } a \qquad \text{meas. of } a$$

$$\psi \quad \rightarrow \quad \psi' \quad \rightarrow \quad \psi'. \tag{7}$$

The first step in (7) can be represented more conveniently simply by replacing ψ' by the identical symbol

$$\mathbf{a}\psi \equiv \psi'. \tag{8}$$

Notice that $\mathbf{a}\psi$ means exactly the same as ψ' but provides a more explicit notation: the prime in ψ' is used to indicate that ψ' has been obtained by some modification of ψ. The symbol $\mathbf{a}\psi$ states exactly this, but it also tells us what has modified ψ: the measurement of a. The symbol $\mathbf{a}$, therefore, is a more explicit modifier than the prime, and is called the *operator* that corresponds to the variable a.

In the same manner, the second step in (7) can be represented with

$$\mathbf{a}\psi' = \psi'. \tag{9}$$

This means that ψ' is a state that is unperturbed by a measurement of the variable a (that is, as we can see from Fig. 4, a state for which a

---

† The reader may be baffled by the curious nature of the measuring instrument represented in Fig. 4, more like a sieve than a ruler. In fact, at the microscopic level measuring instruments must separate out of a complex system particles with a more or less precise value of some desired property. If, for instance, in the figure the electrons were allowed to impinge on a graduated fluorescent screen, the particles in a well defined, measured, state would not be available for further experiments.

has a precise or sharp value). Such a state is called an *eigenstate* of **a** and the corresponding wave function is called an *eigenfunction* of **a**.

If φ is an eigenstate of **a**, it must verify an equation like (9),

$$\mathbf{a}\varphi = \varphi, \tag{10}$$

but we shall see that this is not general enough. In fact, (10) asserts that the state φ is unchanged when we measure a, but this does not guarantee that the resultant state will have exactly φ for state function,

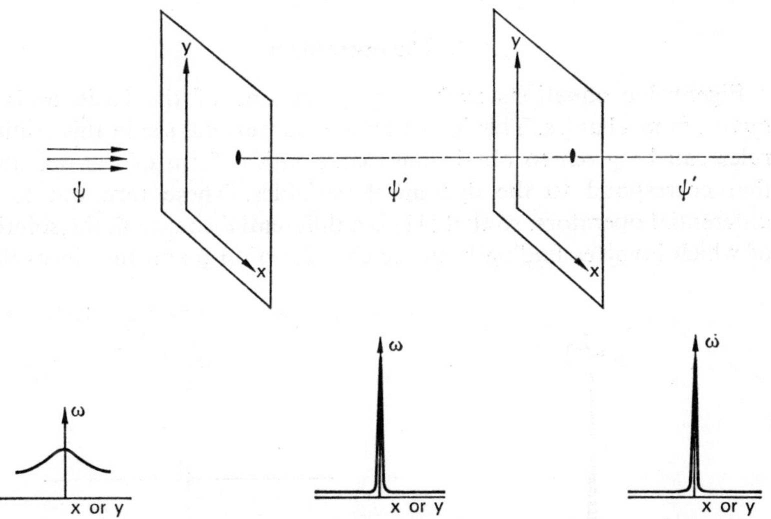

FIG. 4. Repeated measurement of position.

since we have seen (§ 5) that $a'\varphi$ (a' a constant) and φ represent the same state. Therefore, all that we can assert in the right-hand side of (10) is that the resulting state function will be of the form $a'\varphi$:

$$\mathbf{a}\varphi = a'\varphi. \tag{11}$$

When the theory is developed further (see § 17 which, if desired, can be read immediately after this), it is found that a' cannot be an arbitrary constant: it must be the value of the variable a which results from its measurement on the system which has φ for its state function. Such a value a' of a which can be possessed by one of its eigenstates is called an *eigenvalue* and (11) is the *eigenvalue* equation for the variable a. This will often be written in the simplified form $\mathbf{a}\varphi = a\varphi$, in which case it must be remembered that no distinction in notation is made between the variable a and the particular value of it that appears on the right-hand side of this equation.

The treatment of this section has been very formal: the practical uses of this formalism will appear in the next few sections.

7. The operator p

Eigenvalue equations such as (11) are one of the basic tools of quantum mechanics. This is so because, as we shall see in this section, rules can be given to obtain the mathematical form of the operators that correspond to the dynamical variables. These turn out to be differential operators, so that (11) is a differential equation, the solution of which involves finding both the eigenfunction φ and the eigenvalue.

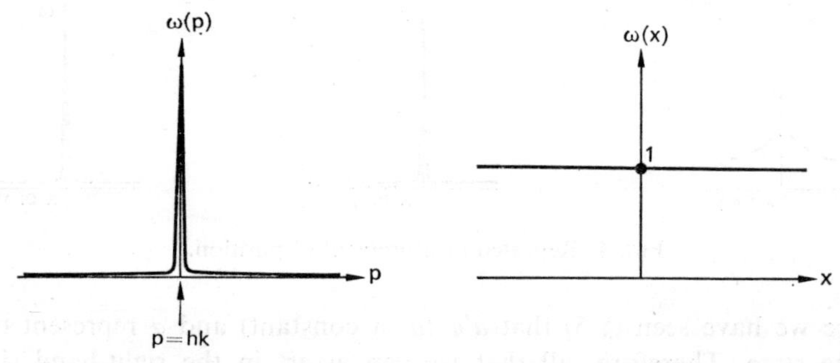

FIG. 5. $\omega(p)$ and $\omega(x)$ for a system with sharp momentum.

In order to illustrate this process we shall take as our dynamical variable the momentum p. We shall make use of the experimental results of Davisson and Germer to guess the form of the corresponding operator $\mathbf{p}$. We have in this experiment a beam of electrons of sharp momentum $p = hk$. Therefore $\Delta p = 0$ whence, from Heisenberg's relation, $\Delta x = \infty$ (which means that all values of the position of the particle are equally probable). We represent the corresponding probability distributions in Fig. 5, where we have taken the constant value of $\omega(x)$ as equal to unity, since it can be readily seen that any other constant value would only introduce a non-significant constant factor in the state function (see § 5).

We know two things about the eigenvalue equation for p:

$$\mathbf{p}\varphi = p\varphi, \tag{12}$$

namely that $p = hk$ and that $\varphi^*(x)\,\varphi(x) = 1$ for all x. This latter condition suggests that $\varphi(x)$ must be of the form $\varphi(x) = \exp(i\alpha x)$, with α a constant (to be determined).

Equation (12) now shows that the operator $\mathbf{p}$ must be such that, when it acts on the function $\exp(i\alpha x)$ it has no other effect on it than multiplying it by a constant. $\mathbf{p}$ must therefore be of the form d/dx or, more generally, $\beta d/dx$ (β constant). All that is now required is to adjust the values of the constants α and β until agreement with the experimental results is obtained. We shall spare the reader the trial and error process involved, and shall simply state the result:

$$\mathbf{p} = \frac{h}{2\pi i}\frac{d}{dx}. \tag{13}$$

In fact, (12) now becomes the differential equation

$$\frac{h}{2\pi i}\frac{d}{dx}\,\varphi(x) = p\varphi(x), \tag{14}$$

the solution of which is

$$\varphi(x) = \exp(2\pi i k x), \tag{15}$$

as can be easily verified on introducing (15) into (14):

$$\frac{h}{2\pi i}\frac{d}{dx}\exp(2\pi i k x) = \frac{h}{2\pi i}2\pi i k \exp(2\pi i k x) = hk\exp(2\pi i k x). \tag{16}$$

Equation (16) shows that (14) is satisfied, for which it is enough to take $p = hk$, in agreement with the experimental result.

We notice that the state function of the electrons in the beam is from (15) a *plane wave* (of wave number k). This is the reason why, as mentioned in § 3, the state function is called the wave function.

We have obtained **p** for free electrons, since in the Davisson and Germer experiment they move in a constant potential field. We shall now *postulate* that the operator (13) is valid even when the potential field is not constant.†

8. The operator **x**

This operator must satisfy the equation

$$\mathbf{x}\varphi(x) = x'\varphi(x), \tag{17}$$

where x' is the position of the particle and $\varphi(x)$ is the eigenfunction that corresponds to this sharp value of x. Clearly, x' will be the only

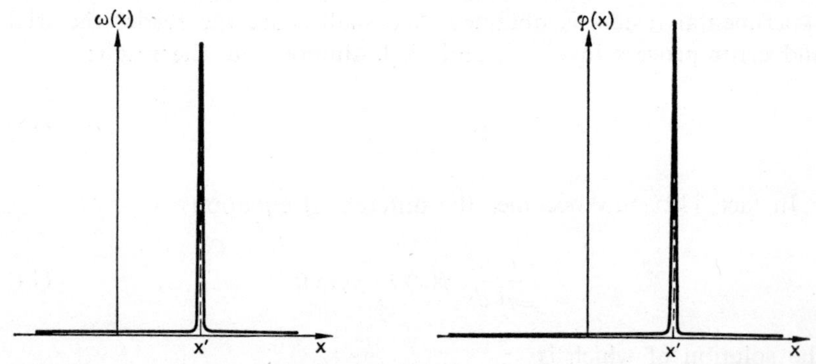

FIG. 6. The probability distribution $\omega(x)$ and the eigenfunction $\varphi(x)$ for a particle with sharp position.

† This postulate can be justified by showing that in the classical limit (that is when quantum effects can be neglected) (13) leads to the correct classical mechanical equations. This important methodological principle is called the *correspondence principle*.

value of x for which $\omega(x)$ is different from zero, and the same must be true for $\varphi(x)$. The form of these functions is shown in Fig. 6. Since

$$\varphi(x) = 0 \quad \text{for} \quad x \neq x',$$

it follows that

$$x\varphi(x) = x'\varphi(x). \tag{18}$$

On comparing (17) and (18), it follows that

$$\mathbf{x}\varphi(x) = x\varphi(x), \tag{19}$$

and hence that the operator $\mathbf{x}$ is equivalent to the prescription "multiply by x". [The reader must beware that although (19) looks like an eigenvalue equation, it is not so: it is for this reason that we made a careful distinction in this section between the variable x and one of its eigenvalues x'.]

9. The Hamiltonian and the Schrödinger equation

The energy can be written in classical mechanics as the sum of the kinetic ($\tfrac{1}{2}\mathbf{m}v^2 = p^2/2\mathbf{m}$) and potential [$V(x)$] energies. The following form of the energy as a function of p and x,

$$H = p^2/2\mathbf{m} + V(x), \tag{20}$$

is called in classical mechanics the hamiltonian function. In quantum mechanics, in the same manner that we associated with p the operator $\mathbf{p}$, we must associate with H an operator $\mathbf{H}$ which is called the *hamiltonian operator*. The rule to write it down, which can be justified by the correspondence principle (§ 7), is very simple: replace p and x in (20) by their corresponding operators $\mathbf{p}$ and $\mathbf{x}$:

$$\mathbf{H} = \mathbf{p}^2/2\mathbf{m} + V(\mathbf{x}). \tag{21}$$

When we introduce in (21) $\mathbf{p} = (h/2\pi i)\, d/dx$ (§ 7) and $\mathbf{x} = x$ (§ 8), we obtain

$$\mathbf{H} = -\frac{h^2}{8\pi^2\mathbf{m}} \frac{d^2}{dx^2} + V(x). \tag{22}$$

The corresponding eigenvalue equation,

$$\mathbf{H}\varphi = E\varphi, \tag{23}$$

in which, to follow tradition, we write E for the eigenvalue of the energy, is called the *Schrödinger equation*.

10. Boundary conditions: quantization

Consider an electron moving in a periodic lattice, such as the linear chain of Fig. 7. If we want to obtain the eigenfunctions of the momentum in this problem, we must solve the eigenvalue equation

$$\mathbf{P}\varphi = p\varphi, \tag{24}$$

with $\mathbf{P} = (h/2\pi i)\, d/dx$. This is a differential equation that must be solved for φ and p. As in § 7,

$$\varphi(x) = \exp(2\pi ikx), \tag{25}$$

$$p = hk. \tag{26}$$

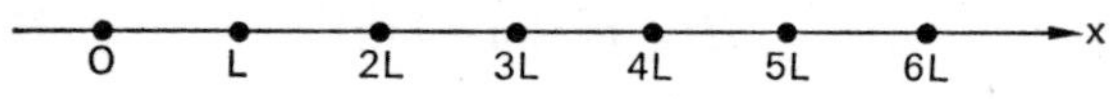

FIG. 7. A periodic lattice.

Because of the periodicity of the problem everything that has a physical meaning for the system must be periodic, and so must be the wave function, which must therefore satisfy the *boundary condition*

$$\varphi(0) = \varphi(L). \tag{27}$$

On introducing (25) into (27), we have $\exp(2\pi ik0) = \exp(2\pi ikL)$, that is,

$$1 = \exp(2\pi ikL). \tag{28}$$

(Remember that $\exp(0) = \cos 0 + i \sin 0 = 1$.)

From (28), kL must satisfy the condition $kL = $ integer $= \kappa$, say, that is

$$k = \frac{\kappa}{L}, \quad \kappa = 0, \pm 1, \pm 2, \ldots \tag{29}$$

and, correspondingly,

$$p = \frac{h\kappa}{L}, \quad \kappa = 0, \pm 1, \pm 2, \ldots \tag{30}$$

which means that the possible values that the momentum can take in this problem form a discrete set. We say that the momentum is *quantized*.

This is a very important result: if the potential energy of the periodic lattice is small enough to be negligible, $E = p^2/2m$, and it follows that the energy of the electrons that propagate in the lattice is also quantized, whereas classically the energy can vary continuously. The existence of discrete levels of the energy in atoms was first postulated by Niels Bohr (1913), and was one of the strange features of the now called "old quantum mechanics". It should be noticed that in the present theory quantization appears very naturally as a direct consequence of the boundary conditions imposed on a given problem.

From (30) any two successive eigenvalues of **p** are separated by the quantity h/L. If L is very large with respect to h (i.e. of macroscopic dimensions such as 10^{-3} cm, say) then the discrete distribution of eigenvalues of **p** is such that two contiguous eigenvalues are separated by an infinitesimal quantity. In such a case, the discrete distribution is called a *quasi-continuous distribution* of eigenvalues.

11. Degeneracy

Consider the eigenvalue equation of an operator **a**,

$$\mathbf{a}\varphi = a\varphi. \tag{31}$$

The eigenfunction φ that corresponds to the eigenvalue a is not unique: any function $c\varphi$ (c a constant) is also an eigenfunction of **a** corresponding to the same eigenvalue a. In fact:

$$\mathbf{a}(c\varphi) = c\mathbf{a}\varphi = ca\varphi = a(c\varphi). \tag{32}$$

(In the first step here, we recognize that since c is a constant it can be taken out of the operand $c\varphi$; in the second we use (31). The third step is a trivial rearrangement of constants: the brackets are not significant but have been added for convenience. Comparison of (32) with (31) demonstrates the result.)

Suppose now that (31) admits of an eigenfunction φ' that corresponds to the eigenvalue a† but which nevertheless is not a multiple of $\varphi : \varphi' \neq c\varphi$ (we say that φ and φ' are *independent*). If this is the case the functions φ and φ' are said to be *degenerate* eigenfunctions of (31).

More generally, if $\varphi_1, \varphi_2, \ldots, \varphi_n$ are n *independent* eigenfunctions of the same operator that correspond to the same eigenvalue of it, they are said to be *n-fold degenerate*.

PROBLEM. Consider two degenerate eigenfunctions φ_1 and φ_2:

$$\mathbf{a}\varphi_1 = a\varphi_1, \tag{33}$$

$$\mathbf{a}\varphi_2 = a\varphi_2. \tag{34}$$

Prove that the function

$$\varphi = c_1\,\varphi_1 + c_2\,\varphi_2 \tag{35}$$

is also an eigenfunction of $\mathbf{a}$ corresponding to the eigenvalue a. [A function given by (35) is said to be *linearly dependent* on φ_1 and φ_2.]

Hint. Use the fact that $\mathbf{a}(f+g) = \mathbf{a}f + \mathbf{a}g$, which is evident for an operator such as d/dx. (Operators that satisfy this condition, as well as $\mathbf{a}(c\varphi) = c\mathbf{a}\varphi$, are called *linear operators*. Most operators used in quantum mechanics must be linear.)

12. Commuting operators

The product of two operators $\mathbf{ab}$ is defined by their successive application, from right to left, on an operand. This product may not be commutative, i.e. it may be that

$$\mathbf{ab} \neq \mathbf{ba}. \tag{36}$$

† Often, rather than saying that a function corresponds to a given eigenvalue, it is said that it *belongs* to it.

EXAMPLE. To prove that $\mathbf{px} \neq \mathbf{xp}$.

Apply the operator $\mathbf{px}$ on a function $\varphi = \varphi(x)$:

$$\mathbf{px}\varphi = \frac{h}{2\pi i}\frac{d}{dx}\,x\varphi = \frac{hx}{2\pi i}\frac{d\varphi}{dx} + \frac{h}{2\pi i}\varphi.$$

In a similar manner,

$$\mathbf{xp}\varphi = \mathbf{x}\frac{h}{2\pi i}\frac{d\varphi}{dx} = x\frac{h}{2\pi i}\frac{d\varphi}{dx}.$$

THEOREM. *If two operators commute, an eigenfunction of one is an eigenfunction of the other and vice versa.*

We shall prove this theorem in the restrictive case of no degeneracy, since the proof for the general case would take us too far. We have to prove that, if

$$\mathbf{a}\varphi = a\varphi \tag{37}$$

and

$$\mathbf{ab} = \mathbf{ba}, \tag{38}$$

then φ is an eigenfunction of $\mathbf{b}$. [Since we have assumed no degeneracy we know that any other eigenfunction of (37) must be a multiple of φ.]

Apply $\mathbf{b}$ on both sides of (37):

$$\mathbf{ba}\varphi = a\mathbf{b}\varphi. \tag{39}$$

From (38),

$$\mathbf{ab}\varphi = a\mathbf{b}\varphi, \tag{40}$$

which we can rewrite as $\mathbf{a}(\mathbf{b}\varphi) = a(\mathbf{b}\varphi)$, to recognize that $\mathbf{b}\varphi$ is an eigenfunction of (37) corresponding to the eigenvalue a. Therefore, it must be a multiple of φ:

$$\mathbf{b}\varphi = b\varphi, \tag{41}$$

say, which proves the theorem.

This result suggests the physical meaning of commutation: if two operators $\mathbf{a}$ and $\mathbf{b}$ commute they have the same eigenfunctions. Therefore, a given system can be simultaneously in an eigenstate of $\mathbf{a}$ and $\mathbf{b}$ so that the variables a and b can simultaneously have sharp values in the given system or, in other words, they can simultaneously be measured exactly. Contrariwise, if **a** and **b** do not commute, the dispersions Δa and Δb of the corresponding variables are connected by a Heisenberg uncertainty relation of the form $\Delta a\, \Delta b = h$, so that a and b cannot have simultaneously sharp values in the same system.

13. Physical meaning of degeneracy

Let us consider a free electron of mass **m**, for which the hamiltonian is

$$\mathbf{H} = \mathbf{p}^2/2\mathbf{m}, \tag{42}$$

since $V(\mathbf{x}) = 0$ (cf. 21). It is clear that **H** and **p** now commute:

$$\mathbf{Hp} = \mathbf{p}^3/2\mathbf{m} = \mathbf{pH}, \tag{43}$$

which means that the eigenfunctions (15) of the momentum are also the eigenfunctions of **H** for a free electron.†

We represent in Fig. 8a the electron propagating to the right in free space so that its momentum is positive with respect to the axis conventionally chosen. However, everything is symmetrical in free space with respect to the point O (which is called a *centre of inversion*), so that

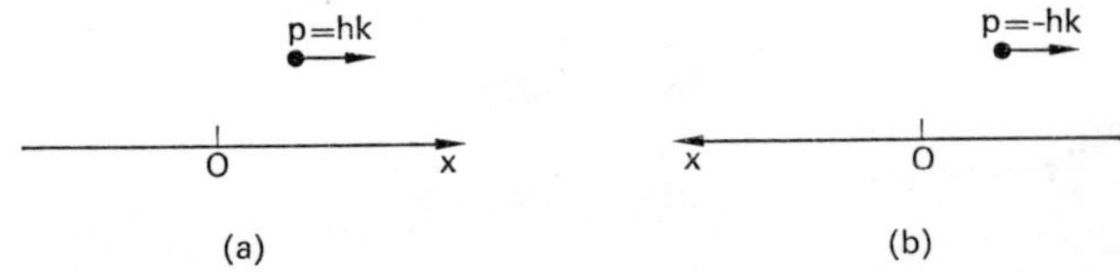

FIG. 8. Effect of the inversion on the momentum.

we can invert the x axis as shown in Fig. 8b. In spite of the symmetry of the system, we cannot expect that everything will remain invariant with respect to this change of axis: this will be so for scalar quantities (like the energy), but not for vectorial ones, like velocities and momenta. In fact, we see in Fig. 8b that the momentum changes sign.

† This result illustrates the usefulness of the theorem of § 12, which often allows us to obtain the eigenfunctions of the hamiltonian from those of much simpler operators that commute with it.

Let us now see what happens to the eigenfunctions:

$$\text{In (a):} \quad p = hk, \quad \varphi_k = \exp(2\pi ikx). \tag{44}$$

$$\text{In (b):} \quad p = -hk, \quad \varphi_{-k} = \exp(-2\pi ikx). \tag{45}$$

The wave function in (44) is straightaway that given in § 7, and the reader can check that the wave function in (45) gives correctly the eigenvalue $p = -hk$ when fed into the eigenvalue equation (24). (More quickly, it can be noticed that, in going from a to b, k is changed into $-k$.)

Since $\mathbf{p}$ and $\mathbf{H}$ commute, φ_k and φ_{-k} are eigenfunctions of $\mathbf{H}$. In order to find their corresponding energy eigenvalues E_k and E_{-k}, it is enough to replace $\mathbf{p}$ in (42) by the respective values in (44) and (45). It follows that $E_k = E_{-k} = h^2 k^2/2\mathbf{m}$, which shows that φ_k and φ_{-k} are degenerate eigenfunctions of $\mathbf{H}$. (Notice incidentally that, as expected, the change of axes effected has left the eigenvalues of the energy invariant.)

This example illustrates the origin of degeneracy. We have eigenfunctions of a vectorial operator $\mathbf{a}$ (like $\mathbf{p}$ in our case) and we change our axes under an allowed *symmetry operation*. Since $\mathbf{a}$ is vectorial its eigenfunctions change (just as φ_k changes into φ_{-k}). If $\mathbf{a}$ commutes with $\mathbf{H}$, all the eigenfunctions that arise are eigenfunctions of $\mathbf{H}$ that belong to the same eigenvalue of it (since $\mathbf{H}$ is scalar). Furthermore, just as φ_{-k} can be obtained by the symmetry operation $x \rightarrow -x$ applied on φ_k, so can all these degenerate eigenfunctions be obtained from one another by some symmetry operation of the system.

Such a relation between the eigenfunctions of a degenerate set is the one that normally exists, although more rarely one finds functions that are degenerate and yet are not related by symmetry. Such a case is called *accidental degeneracy*.

14. Exercise: electron in a box

Consider a one-dimensional box of length L (Fig. 9), where the potential is zero inside the box and infinite outside it. (This means that the probability of finding the electron outside the box is zero, since it would have to surmount an infinite potential barrier.) The boundary conditions, therefore, are

$$\varphi(0) = \varphi(L) = 0 \tag{46}$$

[compare with the boundary conditions (27)].

FIG. 9. One-dimensional box.

Show that the wave functions are

$$\varphi(x) = \sin 2\pi kx, \tag{47}$$

where $k = \kappa/2L$ and $\kappa = 1, 2 \ldots$

Plot the first few eigenfunctions and probability distributions as a function of x.

Hint. The free-electron eigenfunctions $\exp(2\pi ikx)$ cannot vanish for $x = 0$. However, recognize that $\exp(2\pi ikx)$ and $\exp(-2\pi ikx)$ are degenerate and that therefore you can form any combination of them. (See the Problem in § 11.) Notice that if you choose the combination $\exp(2\pi ikx) - \exp(-2\pi ikx)$, the sine functions that appear satisfy the boundary conditions.

Notes. (i) The sine functions (47) represent *standing waves* since they have fixed nodes (at 0 and L), whereas the exponential functions represent *travelling waves.*

(ii) Observe that the boundary condition (46) alters the wave function but superficially with respect to the wave function given by the periodic boundary condition (27). In fact, the sine functions, as mentioned in the hint, are implicit in the set of exponential functions obtained with periodic boundary conditions.

(iii) Observe that, disregarding the nodes at the ends, $|\kappa| - 1$ gives the number of nodes of $\varphi(x)$. Since $E = h^2 k^2/2\mathbf{m}$ and k is proportional to κ, it follows that the higher the number of nodes of a wave function, the higher the energy eigenvalue that corresponds to the wave function. Although this result is strictly valid for free electrons only, it still gives a rough indication about the order of the energy eigenvalues for electrons that move in weak potential fields.

15. Time dependence† (→16)

We defined a stationary state as one for which the probability distribution is independent of the time. It must be appreciated, however, that the corresponding wave function will in general be time dependent. In fact, if we now write Ψ for the wave function this can have the form

$$\Psi(x, t) = \psi(x)\exp(i\alpha t)$$

and yet give place to a time-independent probability distribution, since

$$\omega = \Psi^*\Psi = \psi^*(x)\exp(-i\alpha t)\,\psi(x)\exp(i\alpha t) = \psi^*(x)\,\psi(x),$$

which is a function of x only.

In order to find the time dependence of such a wave function the constant α must be determined, for which we require the *Schrödinger equation with the time* (see § 3), which we quote without proof:

$$\mathbf{H}\Psi = -\frac{h}{2\pi i}\frac{\partial}{\partial t}\Psi. \tag{48}$$

When we introduce $\Psi = \psi(x)\exp(i\alpha t)$ into (48), we have

$$\mathbf{H}\Psi = -\frac{h}{2\pi i}\psi(x)i\alpha\exp(i\alpha t) = -\frac{h}{2\pi}\alpha\Psi, \tag{49}$$

which, on comparing with the Schrödinger equation for a stationary state $\mathbf{H}\Psi = E\Psi$, gives $\alpha = -2\pi E/h$. Hence, the more general wave function for a stationary state is

$$\Psi(x, t) = \psi(x)\exp\left(-\frac{2\pi i}{h}Et\right), \tag{50}$$

an expression which will be required in the next section.

† The contents of this and the following sections of this chapter will not be required until Chapter 3. In a first reading, the reader may proceed directly to Chapter 2.

Before we leave this subject, however, it is useful to go back to the Schrödinger equation with the time and to observe that it actually does the job that was required in § 3, namely to predict the value of the wave function at the time t' from its known value at the time t. This is so because (48) is a differential equation of the first order in t, which, on integration, will provide Ψ at any time in terms of a constant, which is found in terms of the initial value of Ψ.

16. Wave and group velocities ($\rightarrow$ 3.16)

The reader will have noticed that right from the beginning we departed from the normal practice in elementary mechanics whereby the state of a particle is described by its position x and its velocity v. We have used the momentum p instead of the latter, and it is time that we explained why we have done this, which is not a trivial change of notation as it might at first appear. We shall see that we must be very careful because, in quantum mechanics, the velocity is not always given as $v = p/m$.

To start with, we must recognize that there are now two velocities to be considered. This is so because a particle of momentum p (equal to $\hbar k$ from the de Broglie relation) has associated with it a wave of frequency ν. We might think that the velocity of the wave v_p should be the same as that of the particle v_g, but this is not so. (It will soon become clear why we use the suffices p and g to denote these two velocities.)

In fact, it can be quickly seen that we must expect v_p and v_g to be different. It is well known from the elementary theory of wave motion that the velocity v_p of a wave is given in terms of its frequency ν and wavelength λ by

$$v_p = \lambda \nu. \tag{51}$$

Introduce this value into Planck's relation,

$$E = h\nu = h\frac{v_p}{\lambda} = hv_p k = pv_p \tag{52}$$

(replace $1/\lambda$ by k and, in the last step, use $p = \hbar k$). Consider now the case of a free particle for which $E = p^2/2m$ and introduce this value in (52). Then

$$v_p = \frac{p}{2m}, \tag{53}$$

which is, of course, different from the classical value of the velocity as p/m. This shows at once that we must be careful when dealing with velocities. Furthermore, we may expect that at least for a free particle p/m still represents correctly its particle velocity v_g (which will be shown to be true at the end of this section) so that v_p and v_g must be expected to be different.

We must now try to define more precisely what we mean by wave and particle velocities, which are called *phase velocity* v_p and *group velocity* v_g respectively.

We first require a more detailed form of the wave function. Let us introduce in (50) the free-electron wave function $\psi(x) = \exp(2\pi i k x)$, from (25). Then

$$\Psi = \exp(2\pi i k x)\exp\left(-\frac{2\pi i}{h}Et\right), \tag{54}$$

which, on writing $E = h\nu$, becomes

$$\Psi = \exp[2\pi i(kx - \nu t)]. \tag{55}$$

We recognize, as in § 14, that the function obtained from (55) by changing the sign of the exponent is degenerate with Ψ. If we call these two functions Ψ_+ and Ψ_- respectively, the combination $\frac{1}{2}\Psi_+ + \frac{1}{2}\Psi_-$ will be degenerate with the two, and it has the form

$$\Psi = \cos 2\pi(kx - \nu t), \tag{56}$$

which coincides with the expression used in the elementary theory of plane waves. We can write (56) more generally as

$$\Psi = A\cos 2\pi(kx - \nu t), \tag{57}$$

where A is the *amplitude* of the wave (remember that a wave function can always be multiplied by a constant). As in the elementary theory, the velocity of this wave or *phase velocity* is given by (51) or, which is the same, by

$$v_p = \nu k^{-1}. \tag{58}$$

In order to define the group velocity we first add up two waves of the same amplitude:

$$\Psi = A \cos 2\pi(kx - vt), \tag{59}$$

$$\Psi' = A \cos 2\pi(k'x - v't). \tag{60}$$

In forming the sum Φ we use the trigonometric relation $\cos(\alpha + \beta) = 2 \cos \frac{1}{2}(\alpha - \beta) \cos \frac{1}{2}(\alpha + \beta)$:

$$\Phi = 2A \cos \pi[(k - k')x - (v - v')t] \cos \pi[(k + k')x - (v + v')t]. \tag{61}$$

If the difference between the two waves is very small, $k - k' = \Delta k$ and $v - v' = \Delta v$ are small and we can write approximately

$$\Phi = [2A \cos \pi(x\Delta k - t\Delta v)] \cos 2\pi(kx - vt). \tag{62}$$

We observe that the resultant wave can be considered as a plane wave just like (59), but one in which the amplitude [the factor in square brackets in (62)] is *modulated*, that is it is itself a plane wave of wave

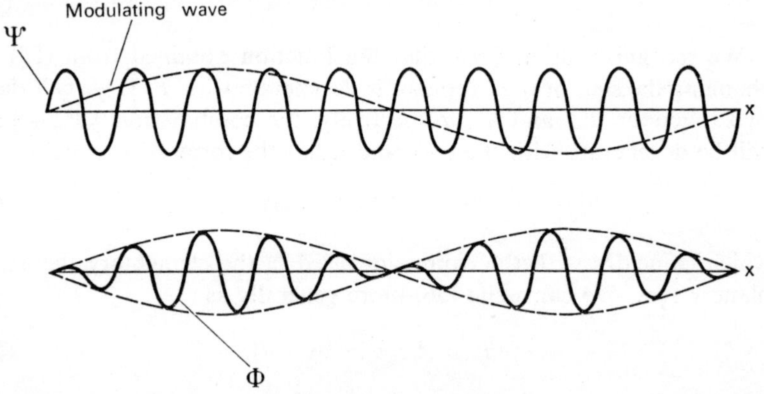

FIG. 10. A modulated wave.

number $\frac{1}{2}\Delta k$. We represent in Fig. 10 the original wave Ψ (which is supposed to be practically indistinguishable from Ψ') and the resultant wave Φ. (In drawing the figure $\frac{1}{2}\Delta k$ was taken to be $\frac{1}{10}k$, so that the wavelength of the modulating wave is ten times that of Ψ. Also, for convenience, their amplitudes are represented as being the same.)

We now define the *group velocity* v_g as the velocity of the envelope of the resultant wave, i.e. of the dotted wave represented in the figure, which we write from (62) as

$$\varphi = 2A\, 2\pi(\tfrac{1}{2}x\Delta k - \tfrac{1}{2}t\Delta v). \tag{63}$$

This is the velocity with which the groups (also called packets) that make up the resultant wave propagate. Applying (58) (understood as: the velocity of the wave equals the coefficient of t divided by the coefficient of x) to the wave (63), we obtain $v_g = \Delta v/\Delta k$, and, on taking the limit for infinitesimal quantities,

$$v_g = dv/dk. \tag{64}$$

In quantum mechanics the group velocity is identified with the velocity of the particle which is associated with the wave. Whereas we cannot fully justify this interpretation, we can give it some plausibility as follows. We notice from (55) that the wave Ψ is completely delocalized since it gives a uniform probability distribution throughout the x axis, so that it does not present anything resembling particle-like behaviour. The wave Φ, on the other hand, is somewhat localized, since there are now condensed regions of space in which the probability of finding the particle is higher. It is essentially these condensed regions that exhibit a particle-like behaviour, and it is the velocity with which these regions travel (i.e. the group velocity) the one that represents the velocity of the particle.

It is convenient to write $v = E/h$ in (64) and then

$$v_g = \frac{1}{h}\frac{dE}{dk}. \tag{65}$$

As an example, for a free particle,

$$E = \frac{p^2}{2m} = \frac{h^2k^2}{2m},$$

whence

$$v_g = \frac{hk}{m} = \frac{p}{m}, \tag{66}$$

in agreement with the classical formula. It must be remembered, however, that, for bound particles, $v_g \neq p/m$ and that therefore the distinction between velocity and momentum is not trivial.

● 17. Mean values ($\rightarrow$ 18)

Consider a variable a which can take discrete values a_x, $(x = 1, 2, \ldots)$ which appear n_x times each in a series of measurements, the total number of which is n. The average value $\bar{a}$ of this variable is, as is well known,

$$\bar{a} = \frac{\sum n_x a_x}{n}, \tag{67}$$

where the summation is over all x. Since $n_x/n = \omega_x$, the frequency or probability of the value a_x,† we have

$$\bar{a} = \sum \omega_x a_x. \tag{68}$$

We must introduce a few simple changes if x is a continuous variable (in fact, from now on, we shall identify it with the coordinate of a particle). Firstly, the values of the variable will now be written, as is usual, $a(x)$ rather than a_x. Secondly, we must substitute the probability density $\omega(x)$ for ω_x (see § 5), and, thirdly, the summation in (68) must be transformed into an integration:

$$\bar{a} = \int \omega(x)\, a(x)\, dx. \tag{69}$$

We now replace $\omega(x)$ by its quantum-mechanical value given by (6), and obtain

$$\bar{a} = \frac{\int \psi^*(x)\, a(x)\, \psi(x)\, dx}{\int \psi^*(x)\, \psi(x)\, dx}. \tag{70}$$

Suppose now that the variable a is a function, not of x but of p, $a(p)$. How do we get its average value? We shall have for it an expression

---

† The symbol $\omega(x)$ has also been used in this chapter (§ 2) to designate this quantity: it will now be used only for the probability density.

formally similar to (70) but with $a(p)$ instead of $a(x)$. However, in order to perform the integration, everything in the integrand must be given in terms of x. Purely formally, we see that we can achieve this result by substituting for $a(p)$

$$a(\mathbf{p}) = a\left(\frac{h}{2\pi i}\frac{d}{dx}\right).$$

We shall more generally write $\mathbf{a}$ for $a(\mathbf{p})$, $\mathbf{a}$ being the operator that corresponds to the variable a, and we shall postulate that the average value of a is given by

$$\bar{a} = \frac{\int \psi^*(x)\,\mathbf{a}\,\psi(x)\,dx}{\int \psi^*(x)\,\psi(x)\,dx}. \tag{71}$$

We can see that this expression is consistent with (70) (since, when a is a function of x, $\mathbf{a} = a(\mathbf{x}) = a(x)$) as well as with our previous theory. In fact, suppose that $\psi(x)$ is now an eigenfunction $\varphi(x)$ of $\mathbf{a}$. Then, from (11),

$$\mathbf{a}\varphi(x) = a'\varphi(x), \tag{72}$$

where, as in (11) a' could be any constant. On introducing (72) in (71), we find that $\bar{a} = a'$. On the other hand, if $\psi(x)$ is an eigenstate of the variable a, the latter must have a (unique) sharp value, which must be repeated for each measurement and for each value of x. Clearly, the average $\bar{a}$ of such a succession of measurements must be the sharp value itself, so that a' in (72) must be equal to the sharp value of the variable a which corresponds to the eigenstate $\varphi(x)$. That is, a' must be the eigenvalue a, as was advanced in § 6.

● 18. The variational method ($\to$ 3.22)

For the particular case of the energy, (71) gives

$$\bar{E} = \frac{\int \psi^*\,\mathbf{H}\,\psi\,d\tau}{\int \psi^*\,\psi\,d\tau}. \tag{73}$$

(In practice, $\mathbf{H}$ and ψ will depend in general on x, y, z or some tri-dimensional coordinates, and the integration must be performed over all space: dx in (71) must be replaced by $d\tau$, the volume element in the system of coordinates used.)

Suppose now that, as is most often the case in practice, ψ is unknown but that nevertheless some approximate value f can be guessed for it. Then we can, in principle, compute

$$\varepsilon \equiv \frac{\int f^* \mathbf{H} f \, d\tau}{\int f^* f \, d\tau}. \tag{74}$$

The following extremely powerful result can be proved in quantum mechanics: as long as f satisfies some simple conditions of behaviour (such as not to have inadmissible singularities), $\varepsilon \geqslant \bar{E}$, the equality sign being valid only when the trial function f coincides with the correct wave function ψ.

This result, which is called the *variational theorem*, permits us to decide, on varying the trial wave function f into another g, whether or not the accuracy of the approximation has improved. To do this, we compute by means of (74) the values of ε that correspond to these two functions, which we shall call ε_f and ε_g respectively. If $\varepsilon_g < \varepsilon_f$, g is a better approximation than f. In fact, in this case, $\varepsilon_f > \varepsilon_g > \bar{E}$, so that ε_g must be nearer to the correct value of the energy than ε_f. This result is the basis of a trial and error method to obtain an approximate wave function, which is called the *variational method*.

It is not difficult to organize systematically this trial and error process. It is enough to make f a function of as many parameters as one can cope with. For each set of numerical values of these parameters a trial wave function is defined, the energy corresponding to which is given by (74), and the parameters are changed until a minimum is obtained. One way in which these parameters are often introduced is as follows. Suppose that we have an initial set of guessed trial functions $f_1, f_2, \ldots, f_n$. These are generally chosen so that they are known to be good approximations to eigenfunctions of the system under study. The functions $f_1, \ldots, f_n$ themselves will not be varied. Instead, a variational function f which involves n variational parameters is formed by writing the sum

$$f = \sum_{i=1}^{n} c_i f_i.$$

The arbitrary coefficients c_i in this expression are the variational parameters to be varied until a minimum of ε is obtained. This form of the trial function is advantageous because it is possible to write explicitly the condition for the minimum, $\partial\varepsilon/\partial c_i = 0$ (all i) as a system of equations from which the energy can be obtained. (See exercise below.)

Exercise

Take a trial wave function

$$f = c_1 f_1 + c_2 f_2$$

and write an expression for the corresponding value of ε in terms of the integrals

$$\left. \begin{aligned} H_{mn} &\equiv \int f_m^* \mathbf{H} f_n \, d\tau, \\[2mm] &\qquad\qquad m, n = 1 \text{ or } 2, \\[2mm] S_{mn} &\equiv \int f_m^* f_n \, d\tau. \end{aligned} \right\} \tag{75}$$

Note. Take $S_{mm} = 1$ (i.e. assume f_1 and f_2 to be normalized).

Method. Write ε from (74) in terms of (74) and (75) as

$$\varepsilon = \frac{c_1^* c_1 H_{11} + c_2^* c_2 H_{22} + c_1^* c_2 H_{12} + c_1 c_2^* H_{21}}{c_1^* c_1 + c_2^* c_2 + c_1^* c_2 S_{12} + c_1 c_2^* S_{21}}. \tag{76}$$

Form $\partial\varepsilon/\partial c_1^*$ to find the condition

$$\frac{c_1 H_{11} + c_2 H_{12} - (c_1 + c_2 S_{12})\,\varepsilon}{c_1^* c_1 + c_2^* c_2 + c_1^* c_2 S_{12} + c_1 c_2^* S_{21}} = 0.$$

On equating the numerator to zero, one equation in c_1 and c_2 is obtained. The equation that follows from the remaining condition, $\partial\varepsilon/\partial c_2^* = 0$, is obtained straightaway by interchanging 1 and 2 in the first equation. You will then have a system of two homogeneous equations in c_1 and c_2, the compatibility condition of which is that the determinant of the coefficients vanishes. This condition, which is called the *secular equation*, is

$$\begin{vmatrix} H_{11} - \varepsilon & H_{12} - \varepsilon S_{12} \\ H_{21} - \varepsilon S_{21} & H_{22} - \varepsilon \end{vmatrix} = 0. \tag{77}$$

This is a quadratic equation that provides ε (two values) in terms of the given integrals. The lowest value of ε corresponds to the ground state of the system under consideration and the other to an excited state of it.

Remarks. (i) The conditions $\partial \varepsilon / \partial c_1 = 0$, $\partial \varepsilon / \partial c_2 = 0$ can readily be seen to yield an equation like (77) in which H_{12}, S_{12}, H_{21}, and S_{21} are replaced by their conjugates. It follows from a general property of the hamiltonian (see Exercise 4, § 19) that H_{11} and H_{22} are real. Also ε, of course, is real. The new equation, therefore, is the complex conjugate of (77) and, since its roots are real, they must coincide with those of (77).

(ii) The most general case, when f_1 and f_2 in the expression $f = c_1 f_1 + c_2 f_2$ are complex, has been considered in this exercise. The variational function f contains therefore four functions and the most general combination of them should contain four coefficients. However, since we have taken c_1 and c_2 to be complex, we had exactly this number of coefficients. In the exercise, we had really four independent variables to vary, namely the real and imaginary parts of c_1 and c_2. It is simpler, however, as we have done, to replace these four variables by c_1, c_2, c_1^*, and c_2^*, treated as independent variables.

(iii) For the general case when $f = \sum_{i=1}^{n} c_i f_i$, it can be proved in the same manner that the condition

$$\det |H_{ij} - \varepsilon S_{ij}| = 0 \tag{78}$$

provides approximations to the n lowest eigenvalues of the system. [Notice that in (78) we give the form of only one element of the determinant, namely that in the ith row and jth column. Verify that (77) is immediately obtained from (78), on remembering that $S_{11} = S_{22} = 1$.]

● 19. Exercises: the hermitian property. Orthogonality
$$(18\leftarrow , \rightarrow 3.22)$$

An operator $\mathbf{R}$ is said to be *hermitian* if the following equality is satisfied for all functions f and g on which it operates:

$$\int f^*(\mathbf{R}g)\,d\tau = \int (\mathbf{R}f)^* g\,d\tau. \tag{79}$$

Exercise 1

Prove that the eigenvalues of a hermitian operator are real.

Method. Apply (79) for the particular case when $f = g$ and g is an eigenfunction of $\mathbf{R}$: $\mathbf{R}g = \varepsilon g$. ε can be taken out of the integral in the left-hand side of (79). On the right-hand side we have $(\mathbf{R}g)^* = \varepsilon^* g^*$ and ε^* is taken out of the integral. Cancelling out the remaining integral on both sides you will obtain $\varepsilon = \varepsilon^*$, whence ε must be real. (Remember that if $\varepsilon = a+bi$, the condition $\varepsilon = \varepsilon^*$ entails $a+bi = a-bi$, that is $b = -b$ and therefore $b = 0$.)

Exercise 2

Consider two eigenfunctions of the same hermitian operator that belong to two different eigenvalues:

$$\mathbf{R}f_m = \varepsilon_m f_m, \tag{80}$$

$$\mathbf{R}f_n = \varepsilon_n f_n, \qquad \varepsilon_m \neq \varepsilon_n. \tag{81}$$

Prove that

$$\int f_m^* f_n\,d\tau = 0. \tag{82}$$

Two functions that satisfy this property are said to be *orthogonal*.

Method. On account of the hermitian property,

$$\int f_m^*(\mathbf{R}f_n)\,d\tau = \int (\mathbf{R}f_m)^* f_n\,d\tau. \tag{83}$$

From (81), take ε_n out of the integral on the left-hand side and, from the complex conjugate equation of (80), ε_m^* on the right. From exercise (1) $\varepsilon_m^* = \varepsilon_m$. Then

$$\varepsilon_m \int f_m^* f_n \, d\tau = \varepsilon_n \int f_m^* f_n \, d\tau.$$

Since $\varepsilon_m \neq \varepsilon_n$ the integral must vanish.

Remark. The reader may puzzle how we got the curious condition (79). He must realize that we have looked at the problem through the wrong end of the telescope. We know that quantum mechanical operators must have real eigenvalues. Also, from some general principles that we have not discussed, it follows that their eigenfunctions must be orthogonal. (An idea of the meaning of this requirement will be obtained from Exercise 3.) A long process of trial and error has shown that the simplest way in which these requirements are guaranteed is by the condition (79). In fact, all quantum mechanical operators must be hermitian.

Exercise 3

Prove that if f_m and f_n are orthogonal they must be independent (see § 11).

Method. You must prove that $f_m \neq c f_n$, where c is a constant different from zero. Multiply both sides of this expression by f_n^* and integrate. If the equality sign were to hold you would have

$$c = \int f_n^* f_m \, d\tau \Big/ \int f_n^* f_n \, d\tau = 0.$$

Remarks. (i) This result shows why the orthogonality condition is important: it is equivalent to stating that eigenfunctions of hermitian operators that belong to different eigenvalues are independent.

(ii) If, in the variational procedure of the exercise in § 18 the functions f_1 and f_2 are not independent, the method given there becomes incorrect: c_1 and c_2 are in this case dependent and we cannot differentiate separately with respect to them as we have done. In general, when variational expressions of the form discussed in that section are used, it is essential to guarantee the independence of the functions $f_1, f_2, \ldots, f_n$. From the result of this exercise it follows that the simplest way to secure this is to ensure that these functions are orthogonal. This is another example of the practical importance of the orthogonality condition.

Exercise 4

Show that, with the definition

$$R_{mn} = \int f_m^* \mathbf{R} f_n \, d\tau, \tag{84}$$

$R_{mn} = R_{nm}^*$, when $\mathbf{R}$ is a hermitian operator. Deduce that R_{mm} must be real.

Exercise 5

Show that the operator $\mathbf{p} = \dfrac{h}{2\pi i}\dfrac{d}{dx}$ is hermitian when the functions on which it acts are periodic: $f(0) = f(L)$.

Method. You must prove that

$$\int_0^L f^*(x)\frac{h}{2\pi i}\frac{d}{dx}g(x)\,dx = \int_0^L\left[-\frac{h}{2\pi i}\frac{d}{dx}f^*(x)\right]g(x)\,dx, \tag{85}$$

where the integration is carried out from 0 to L since this is taken to be the only physically significant part of the x axis (see § 2.4 for an instance when this is the case).

Write the left-hand side of (85) as follows:

$$\frac{h}{2\pi i}\int_0^L f^*(x)\,dg(x) = \frac{h}{2\pi i}f(x)g(x)\Big|_0^L - \frac{h}{2\pi i}\int_0^L g(x)\,df^*(x),$$

where, in the right-hand side, integration by parts has been used. On account of the periodicity condition the integrated part vanishes and therefore

$$\int_0^L f^*(x)\frac{h}{2\pi i}\frac{dg(x)}{dx}\,dx = \int_0^L g(x)\left[-\frac{h}{2\pi i}\frac{d}{dx}f^*(x)\right]dx,$$

which agrees with (85).

Remarks. This is a particular case of the general requirement that all quantum mechanical operators must be hermitian and it can be extended for functions with less restrictive properties than those stated. (Try, for instance, functions that vanish at infinity.) Also notice that (85) and therefore the basic condition (79) are intimately connected with the familiar integration-by-parts method. The hermitian condition (79) is a sort of an integration-by-parts statement, an understanding of which fact helps to remove some of the mystery surrounding the condition.

Exercise 6

$f(x)$ and $g(x)$ are real functions of x. $f(x)$ is *even* $[f(x) = f(-x)]$ and $g(x)$ is *odd* $[g(x) = -g(-x)]$. Prove that $f(x)$ and $g(x)$ are orthogonal in the symmetrical interval $-a \leqslant x \leqslant a$.

Method. You must prove that, if

$$I = \int_{-a}^{a} f(x)\,g(x)\,dx, \tag{86}$$

then $I = 0$. Change the variable x by $-x$ in (86) (remember the limits) and replace $f(-x)$, $g(-x)$ by the relations given for these functions. You will find $I = -I$ whence the result follows.

CHAPTER 2

Free-electron Theory of Metals

We shall introduce in §§ 1–3 of this chapter some approximations that will allow us to construct a model of a metal for which some very simple calculations can be performed. Each approximation should properly be justified by an order of magnitude calculation of the error involved, as was done in some instances when the corresponding theories were first presented. We shall not attempt to do this: instead, we shall describe the approximations required as precisely as possible and shall provide an *a posteriori* justification of the model through the results of the calculations to be performed in the rest of this chapter.

It will be seen, nevertheless, that the crude model used here provides an understanding of some important experimental facts. Other phenomena lie outside the scope of the free-electron model and will require the more precise theory of Chapter 3, but the free-electron model remains a useful first approximation in many cases.

1. The one-particle approximation

It should be clearly understood that the approximations to be introduced in this and the following section are required not just for the free-electron model: they will remain in use throughout this book.

Suppose that we have inside a very large container two particles that are so far apart that they do not interact at all. Clearly, we can set up a Schrödinger equation for each particle as if the other one did not exist:

$$\mathbf{H}(1)\,\varphi(1) = E_1\,\varphi(1), \tag{1}$$

$$\mathbf{H}(2)\,\varphi(2) = E_2\,\varphi(2). \tag{2}$$

37

The symbols (1) and (2) in these equations stand for the coordinates of the particles on which the hamiltonians and wave functions depend.

Consider now the total system, made up of the two particles. Classically, its energy will be the sum of terms that belong to each particle independently. Correspondingly, the total hamiltonian, which we shall denote with $\mathbf{H}(1,2)$, will be the sum of the hamiltonians of each particle:

$$\mathbf{H}(1,2) = \mathbf{H}(1) + \mathbf{H}(2), \tag{3}$$

and the total energy E will be

$$E = E_1 + E_2. \tag{4}$$

We want to find the total wave function of the composite system, which we shall denote with $\Phi(1,2)$. We assert that this is

$$\Phi(1,2) = \varphi(1)\,\varphi(2). \tag{5}$$

Proof. In order to verify (5) we must prove that

$$\mathbf{H}(1,2)\,\Phi(1,2) = E\Phi(1,2), \tag{6}$$

with E given by (4). In fact,

$$[\mathbf{H}(1) + \mathbf{H}(2)]\,\varphi(1)\,\varphi(2) = \mathbf{H}(1)\,\varphi(1)\,\varphi(2) + \mathbf{H}(2)\,\varphi(1)\,\varphi(2). \tag{7}$$

We must observe that $\mathbf{H}(1)$ and $\mathbf{H}(2)$ act only on $\varphi(1)$ and $\varphi(2)$ respectively: $\varphi(1)$, for instance, is a constant with respect to $\mathbf{H}(2)$, since the latter contains only terms such as "differentiate with respect to coordinates (1)". $\mathbf{H}(1)\varphi(1)$ and $\mathbf{H}(2)\varphi(2)$ are obtained from (1) and (2) respectively and then, from (7),

$$[\mathbf{H}(1) + \mathbf{H}(2)]\,\varphi(1)\,\varphi(2) = E_1\,\varphi(1)\,\varphi(2) + E_2\,\varphi(1)\,\varphi(2)$$

$$= (E_1 + E_2)\,\varphi(1)\,\varphi(2), \tag{8}$$

which agrees with (6).

Similarly, it can readily be seen that, if we have n non-interacting particles, then

$$\mathbf{H} = \mathbf{H}(1) + \mathbf{H}(2) + \ldots + \mathbf{H}(n), \tag{9}$$

and
$$\Phi = \varphi(1)\,\varphi(2)\ldots\varphi(n), \tag{10}$$

where
$$\mathbf{H}(i)\,\varphi(i) = E_i\,\varphi(i), \quad i = 1, 2, \ldots n. \tag{11}$$

Notice that in this case the equation $\mathbf{H}\Phi = E\Phi$ for the total system reduces to the solution of n Schrödinger equations, each of which corresponds to a single particle.

Consider now a system of n interacting particles. We must solve a Schrödinger equation

$$\mathbf{H}(1, 2, \ldots, n)\,\Phi(1, 2, \ldots, n) = E\Phi(1, 2, \ldots, n). \tag{12}$$

The exact solution of (12) is not difficult: it is impossible, even for small values of n. In order to see what can be done about this we shall go back to a two-particle system and now write its hamiltonian as

$$\mathbf{H}(1, 2) = \mathbf{H}(1) + \mathbf{H}(2) + \mathbf{H}'(1, 2). \tag{13}$$

The first two terms here are the same as those in (3) and $\mathbf{H}'(1, 2)$ contains the *interaction* terms, i.e. those that depend on the coordinates of both particles at the same time. For instance, if we have two electrons, the force between them is $-\mathbf{e}^2/r_{12}^2$ ($\mathbf{e}$ = electron charge, r_{12} = distance between the two electrons), which leads to a well-known *Coulomb potential* term in the energy, of the form $\mathbf{e}^2/r_{12}$. This classical term will lead to a similar one in the quantum mechanical hamiltonian, which will appear in $\mathbf{H}'(1, 2)$.

In order to simplify the hamiltonian (13), we first split the interaction term between the two particles, so that we have $\mathbf{H}(1) + \frac{1}{2}\mathbf{H}'(1, 2)$ as the contribution to the hamiltonian coming from the first particle, and a similar term (with $\mathbf{H}(2)$ substituted for $\mathbf{H}(1)$) from the second. $\mathbf{H}'(1, 2)$ depends on the coordinates of particle 1 and those of particle 2. If we can guess an approximate wave function for particle 2 on its own, we obtain an approximate coordinate distribution for particle 2, and we can construct an average for its position. When this average is introduced in $\mathbf{H}'(1, 2)$ we obtain a term that we shall write as $\mathbf{H}'(1, \bar{2})$, which depends only on the coordinates of particle 1, since $\bar{2}$ is of course no longer a variable but a constant average value that appears simply as a parameter in this term. The contribution to the hamiltonian arising from the first particle can now be written as

$$^e\mathbf{H}(1) = \mathbf{H}(1) + \tfrac{1}{2}\mathbf{H}'(1,\bar{2}), \tag{14}$$

and depends exclusively on the coordinates of particle 1. The corresponding term for particle (2) will have the form

$$\mathbf{H}(2) + \tfrac{1}{2}\mathbf{H}'(\bar{1},2).$$

As an approximation, we now accept that (13) can be written as

$$\mathbf{H}(1,2) \simeq {}^e\mathbf{H}(1) + {}^e\mathbf{H}(2), \tag{15}$$

where $\simeq$ means approximately equal. The partial, approximate, hamiltonians that appear on the right-hand side of this equation are called *effective one-particle hamiltonians*. In the same manner, the hamiltonian in (12) can be approximated as

$$\mathbf{H}(1,2,\ldots,n) = {}^e\mathbf{H}(1) + {}^e\mathbf{H}(2) + \ldots + {}^e\mathbf{H}(n), \tag{16}$$

where in $^e\mathbf{H}(i)$ the coordinates of all the particles except the ith one are replaced by some average. With this approximation, the hamiltonian takes the same form as that for n non-interacting particles (9), and its solution, exactly as in that case, will reduce to that of the n one-particle Schrödinger equations

$$^e\mathbf{H}(i)\,\varphi(i) = E_i\varphi(i), \quad i = 1,2,\ldots,n, \tag{17}$$

with an effective hamiltonian.

The difficulty with this procedure is that one must start with some reasonable guess for the one-particle wave functions, in order to construct the required averages. This is not too serious, however, since the process is iterative, so that it can be used as follows. Once we guess $\varphi(2), \varphi(3), \ldots, \varphi(n)$, we can obtain $\varphi(1)$ from (17). This improved value of $\varphi(1)$ can now be used in computing all the others, and the new values of $\varphi(2), \varphi(3), \ldots, \varphi(n)$ thereby obtained will yield an improved value of $\varphi(1)$, and so on. At each stage of the iteration we compute improved values of the effective hamiltonians $^eH(1)$, $^eH(2), \ldots, {}^eH(n)$ and the process continues until these values become stable, on further iteration, to some required order of accuracy. When this is the case, the hamiltonian (16) obtained in the final iteration is called a *self-consistent hamiltonian*, which explains why this procedure, which was first introduced by D. R. Hartree for many-electron atoms, is called the *self-consistent field method*.

In metals, where the computations are much harder, a full iteration is not carried out. It is, nevertheless, important to understand that we must define in the metal an effective single-particle hamiltonian in some approximate manner, that the total hamiltonian is then written as a sum of these "non-interacting" hamiltonians and that the Schrödinger equation reduces therefore to the solution of single-particle Schrödinger equations (17). This, which is called the *one-particle approximation*, although fairly heavily criticized at one time, has received in the last few years a certain amount of support from experimental results, notably through the study of the Fermi surface (see § **6.7.2**).

PROBLEM. The factorization (10) of the wave function is valid not only for the hamiltonian operator. Prove in the manner suggested in the text that if an operator $\mathbf{a}$ can be written in the form $\mathbf{a} = \mathbf{a}(1) + \mathbf{a}(2) + \ldots + \mathbf{a}(n)$, where $1, 2, \ldots, n$ are independent variables, the solution Φ of the equation $\mathbf{a}\Phi = a\Phi$, is of the form $\Phi = \varphi(1)\,\varphi(2)\ldots\varphi(n)$, where $\varphi(i)$ is a solution of the equation $\mathbf{a}(i)\varphi(i) = a_i\varphi(i)$ and $a = a_1 + a_2 + \ldots + a_n$.

2. Core and metal electrons. Pauli principle

Consider, as an example, sodium metal. A free atom of sodium has the configuration $(1s)^2\,(2s)^2\,(2p)^6\,3s$. We expect, of course, the atoms to be somewhat distorted in the metal and, since the ionization potential of the $3s$ electrons is low, and much lower than that of the $2p$ electrons, we shall *assume* that the metal is made up of ion cores at the lattice sites with the configuration $(1s)^2,\,(2s)^2,\,(2p)^6$, very little distorted with respect to the free Na^+ ion, and that the $3s$ electrons are shared by the whole crystal roaming about it more or less freely. We can then assume that it is these outer electrons that contribute to the electron "sea" or "gas" that gives a metal its distinctive properties, which is the virtue of this model. The core electrons, on the other hand, do nothing except contribute to the core potentials, so that in this model the system of many particles that makes up the metal originates from the outer (or valency) electrons of the metal atoms that move in the field of the metal cores. It is for this system of outer or metal electrons that we introduce the one-particle approximation.

When this is done and the one-electron Schrödinger equation is solved, energy levels are obtained which, as we shall see, are quantized, just as those of an electron in a box (see § 1.14). From this point of view the system under consideration behaves in the same manner as an ideal gas of electrons (the word "ideal" indicating here that the particles do not interact). Although the electron system in some ways resembles a molecular gas, there is a very fundamental difference between the two systems, which we illustrate in Fig. 11. The horizontal lines in the figure represent, on a vertical energy scale, the energy levels calculated with the one-particle equation that corresponds to the problem. There are two differences between the molecular and the electron gas. The energy levels for the first are all very crowded together, forming a quasi-continuous distribution, as we might expect, since the classical picture of a molecular gas, in which the energy levels are not quantized at all, is quite a good one. On the other hand, it is not obvious how the energy levels should be distributed in the electron gas in a metal (in Chapter 3, for instance, we shall see that there are regions of the energy scale for which no energy levels exist at all) although, for convenience, they are represented in Fig. 11 as a quasi-continuous distribution as well. The point is that, at this stage, the difference that we want to illustrate is the second one.

This hinges on the way in which the total electron system is built up. The one-particle approximation will give us a set of energy levels which can be occupied by any one particle of the system. Since we have a large number of particles we fill in the energy levels starting at the bottom (just as it is done for a many-electron atom). In a gas, we can put as many particles as we please in each level, so that in principle all the particles will crowd together in the lowest level: this would correspond to the lowest energy state of the gas. In practice, we know from kinetic theory that a gas at the temperature T has an average kinetic energy $E = \frac{3}{2} k_B T$ (k_B = Boltzmann's constant). When the values of the constants are introduced, this gives a value of the order of 0·02 eV at normal room temperature, which means that some of the particles in the gas will have energies up to about 0·02 eV above the lowest level, as shown on the left of the figure. For an electron gas, on the other hand, we must take into account the *Pauli principle*, which

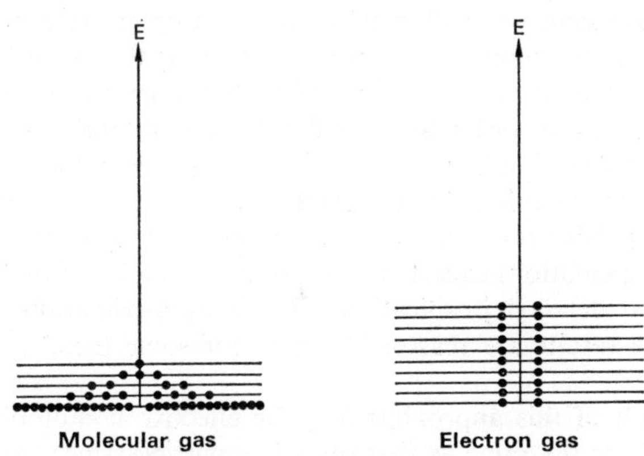

FIG. 11. Occupation of energy states. Each dot represents a particle occupying the corresponding energy state. For the molecular gas, the number of dots in the lowest level should be much greater than shown, of the order of the Avogadro number.

prohibits the occupation of the same state by more than two electrons (of different spins). It should be noticed that the operation of the Pauli principle increases enormously the energy of the electron system, as appears from the figure, and that it gives its very distinctive properties to the electron gas in the metal, which is called the *Fermi gas* (or sometimes the *Fermi sea*), since it was Fermi who first studied the properties of such systems.

It must be carefully remembered that the Pauli principle does not refer to the occupation of an energy level but rather to that of an electron state, each state corresponding, of course, to a wave function. For instance, if we have two degenerate eigenfunctions corresponding to the same energy level, we can put two electrons in each of the two states, which will give a total of four electrons in that particular energy level. Therefore, just as in atoms, a study of the degeneracy of the successive levels is essential in order to obtain a building-up principle for the metal states.

3. The free-electron model

As we have said, we shall consider the electron gas of a metal in the one-particle approximation, i.e. we shall assume that each electron of the gas moves in an averaged field of all the other electrons plus that of the ion cores. In order to write down a hamiltonian and solve the Schrödinger equation for the energy states, we must make some assumption about this field. We shall assume in this chapter that the averaged repulsive potential of all the electrons (except the particular one under calculation) cancels the attractive potential of the ion cores. We offer no detailed justification for this approximation, except to say that the results are reasonably good for some metals, sodium in particular.

As a result of this approximation, the effective hamiltonian for an electron will be the same as that for a free particle, since the potential field that appears in it cancels out. In fact, the n equations (11) for the n electrons in the metal gas are now identical, since all the effective hamiltonians for all the particles in the system are the same. It should be rather pleasing to observe that the initially formidable n electron problem has thus been reduced to the solution of a single one-electron Schrödinger equation.

4. Periodic (Born–von Kármán) boundary conditions

The technique discussed in this section is not peculiar to the free-electron model: in fact it will be used throughout this book.

We shall first discuss a one-dimensional metal and shall leave the consideration of a proper three-dimensional one until later on. Consider a piece of a one-dimensional metal of length L (see Fig. 12, which should be compared with Fig. 9). We ought to apply the boundary conditions of an electron in a box, $\varphi(0) = \varphi(L) = 0$, since we know that a given electron cannot escape out of the metal. We saw, however, in § 1.14 that the box condition does not give a substantial difference with the periodic conditions that apply when the length L is periodically repeated, as in Fig. 13. In particular, the quantization of the energy levels is unchanged.

FIG. 12. A one-dimensional metal.

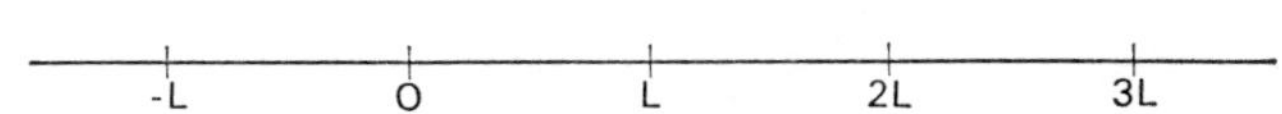

FIG. 13. Periodic repetition of the one-dimensional metal.

Of course, the periodic repetition of the sample of metal under study has no physical meaning. On the other hand, in so far as it represents the sample of metal under study as a part of an infinitely long piece, it should be correct: if the sample of metal is not too small, its properties cannot depend on its size and they should be the same as those of an infinite piece.

Another way in which the periodic boundary conditions can be justified is as follows. If the piece of metal represented in Fig. 12 is long enough we can bend it into a circle of large radius and therefore small curvature (Fig. 14). Since the curvature is small, the properties of any small part of this object must be practically indistinguishable from those of a straight one. On the other hand, it is clear that if we go repeatedly round the circle we satisfy exactly the same periodicity conditions as are valid along the straight line of Fig. 13.

The *periodic boundary conditions*

$$\varphi(0) = \varphi(L) \qquad (18)$$

were first introduced by Born and von Kármán, and we prefer to use them rather than the box conditions, since they allow us to use exponential (or travelling) wave functions which are easier to handle and which include the standing waves of the box as a particular case.

The periodic boundary conditions will be discussed again in Chapter 3.

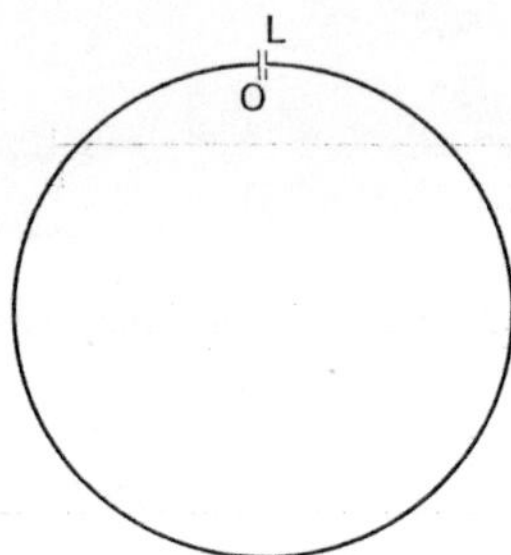

FIG. 14. The metal of Fig. 12 bent into
a circle.

5. The wave function: normalization

In the present picture the electron gas is made up of a large number of particles each of which moves freely in the (zero) field of the other particles plus the ion cores. The effective hamiltonian $^e\mathbf{H}(i)$ for the ith particle therefore describes a free electron. The solution of the one-particle equation (11) with such a hamiltonian can be much simplified since, as we have shown in § **1.13**, the hamiltonian of a free particle commutes with the momentum operator. Therefore, from the theorem of § **1.12**, the eigenfunctions of the momentum are eigenfunctions of the energy. If, for simplicity, we write x for the coordinate of the particle under consideration (rather than x_i, say, for the ith particle) the one-particle wave function $\varphi(x)$ is given straightaway by eqns. **(1.25)** and **(1.29)**:

$$\varphi(x) = \exp(2\pi i k_x x), \tag{19}$$

$$k_x = \kappa_x/L, \quad \kappa_x = 0, \pm 1, \pm 2, \ldots \tag{20}$$

and correspondingly, the energy eigenvalue is

$$E = \frac{p^2}{2m} = \frac{h^2 k_x^2}{2m} = \frac{h^2}{2mL^2}\kappa_x^2. \tag{21}$$

(As a difference with § 1.10 we have introduced here a suffix x in k and κ in order to allow for the consideration of the directions y and z later on.)

We must now introduce a small modification in the wave function so as to normalize it, in order to ensure that the probability of finding the electron in the sample of metal of length L is correctly equal to unity. In fact, $\varphi^*(x)\,\varphi(x)\,dx$ is the probability of finding the electron in a small interval dx around x (see § 1.5). Hence

$$\int_0^L \varphi^*(x)\,\varphi(x)\,dx \tag{22}$$

is the probability of finding the electron in the length L, which we want to be unity. However, with the value of $\varphi(x)$ given by (19),

$$\int_0^L \varphi^*(x)\,\varphi(x)\,dx = \int_0^L \exp\left(-2\pi i k_x x\right)\exp\left(2\pi i k_x x\right)dx = \int_0^L dx = L. \tag{23}$$

Hence, since we can always multiply the wave function by a constant (§ 1.5) we take

$$\psi(x) = L^{-\frac{1}{2}}\exp\left(2\pi i k_x x\right) = L^{-\frac{1}{2}}\exp\left(2\pi i\frac{\kappa_x x}{L}\right), \tag{24}$$

which is now correctly normalized.

● 1. *Exercise: orthonormality of the wave functions* ($\to$ 3.24)

Consider two functions (24) that correspond to two different values of k (i.e. of κ; we shall dispense with the subscript x in this exercise):

$$\psi_i(x) = L^{-\frac{1}{2}}\exp\left(2\pi i\,\kappa_i x/L\right),$$

and, similarly, $\psi_j(x)$. Show that

$$S_{ij} \equiv \int_0^L \psi_i^*(x)\,\psi_j(x)\,dx = 1 \quad \text{if} \quad i = j,$$

$$= 0 \quad \text{if} \quad i \neq j.$$

Method. The first part of the proof, for S_{ii}, is identical, except for a trivial change of notation, with the work of (23) and (24). To prove the second part show that

$$S_{ij} = L^{-1} [2\pi i(\kappa_j - \kappa_i)]^{-1} \left| \exp[2\pi i(\kappa_j - \kappa_i) x/L] \right|_0^L ,$$

and notice that the exponential here is equal to unity when $x = L$ since $\kappa_j - \kappa_i$ must be an integer.

Nomenclature. The result proved shows that the functions ψ_i (all i) are normalized and that any two of them ψ_i and ψ_j (all i and all $j \neq i$) are orthogonal. (See Exercise 2 of § 1.19.) We say that the set of functions $\psi_i(x)$ for all i form an *orthonormal set of functions* for $0 \leqslant x \leqslant L$.

Remark. The orthogonality of $\psi_i(x)$ and $\psi_j(x)$ follows directly from the fact that they are eigenfunctions of a hermitian operator that belong to different eigenvalues of it. (See Exercise 2, § 1.19.)

Notation. In order to be able to give more concisely the properties shown for S_{ij}, the following symbol is defined:

$$\delta_{ij} = 1, \quad \text{when} \quad i = j,$$
$$= 0, \quad \text{when} \quad i \neq j.$$

This symbol is called *Kronecker's delta*, and since its values are universally recognized as those just defined we could have enunciated our theorem as $S_{ij} = \delta_{ij}$.

6. The eigenfunctions in three dimensions

Since the momentum p is a vector, it can be given in three dimensions in terms of its three components along axes x, y, z:

$$\mathbf{p} = \mathbf{p}_x + \mathbf{p}_y + \mathbf{p}_z. \tag{25}$$

Correspondingly, the operator $\mathbf{p}$ will be given by

$$\mathbf{p} = \mathbf{p}_x + \mathbf{p}_y + \mathbf{p}_z, \tag{26}$$

where $\mathbf{p}_x = (h/2\pi i)\,\partial/\partial x$ and similarly for the other components. If, for simplicity, we assume that our metal sample is cubic so that its length has the same value L in the three directions x, y, and z, $\psi(y)$ and $\psi(z)$ will have the same form as (24), with y and z respectively substituted for x. The eigenfunctions $\psi(x, y, z)$ of the operator $\mathbf{p}$ are given, from the result of the Problem in § 1, as a product of the eigenfunctions of $\mathbf{p}_x$, $\mathbf{p}_y$, and $\mathbf{p}_z$:

$$\psi(x, y, z) = \psi(x)\,\psi(y)\,\psi(z)$$

$$= L^{-3/2}\exp\left[2\pi i\,(k_x x + k_y y + k_z z)\right], \tag{27}$$

$$= L^{-3/2}\exp\left[\frac{2\pi i}{L}(\kappa_x x + \kappa_y y + \kappa_z z)\right], \tag{28}$$

which corresponds to the energy eigenvalue

$$E = \frac{h^2}{2mL^2}(\kappa_x^2 + \kappa_y^2 + \kappa_z^2). \tag{29}$$

[cf. (21)].

For simplicity, we shall write

$$\kappa^2 = \kappa_x^2 + \kappa_y^2 + \kappa_z^2, \qquad \kappa_x, \kappa_y, \kappa_z = 0, \pm 1, \pm 2, \ldots \tag{30}$$

and then

$$E = \frac{h^2}{2mL^2}\kappa^2. \tag{31}$$

7. Degeneracy of the levels

Equation (31) provides us with a system of quantized energy levels, similar to the one represented in Fig. 11, which we must fill in starting from the bottom with the particles that make up the electron gas. However, as explained at the end of § 2, it is essential, in order to do this, to know the degeneracy of each level, since if a given level is n-fold degenerate it can accommodate $2n$ electrons.

Consider, for example, the following eigenfunctions [cf. (28), (29), and (30)]:

$$\psi_1 = L^{-3/2}\exp\left(\frac{2\pi i}{L}x\right); \quad \kappa_x = 1, \kappa_y = \kappa_z = 0; \kappa^2 = 1,$$

$$\psi_2 = L^{-3/2}\exp\left(\frac{2\pi i}{L}y\right); \quad \kappa_y = 1, \kappa_y = \kappa_z = 0; \kappa^2 = 1,$$

which are degenerate since κ^2 is the same in both cases and also, therefore (from 31) the energy level E.

It is clear that in order to find the degeneracy of a level we must find all possible combinations of integers $\kappa_x, \kappa_y, \kappa_z$ such that the sum of their squares gives the same value, $\kappa_x^2 + \kappa_y^2 + \kappa_z^2 = \kappa^2$. We do this for the first few energy levels in Table 1.

TABLE 1. DEGENERACIES OF FREE-ELECTRON LEVELS

κ_x	κ_y	κ_z	κ^2	Degeneracy	Degeneracy with spin
0	0	0	0	1	2
± 1	0	0	1		
0	± 1	0	1	6	12
0	0	± 1	1		
± 1	± 1	0	2	4×3	24
± 1	± 1	± 1	3	8	16

Only the combinations that correspond to the first two levels are represented explicitly in the table. In order to count the number of combinations for the third level, notice that there are four ways in which the signs can be combined and three possible positions for the zero.

The results of the table show that the picture of the electron gas given in Fig. 11 should be redrawn as shown in Fig. 15.

The number of electrons in a given metal will depend on various properties of it (number of electrons in the valence shell, lattice constant, etc.). Therefore we shall pile up the electrons in the available levels at the rate of as many electrons per level as permitted by its degeneracy, until all the electrons available in the electron gas of the metal are exhausted, when a top energy level will be reached. (See Fig. 16.) This

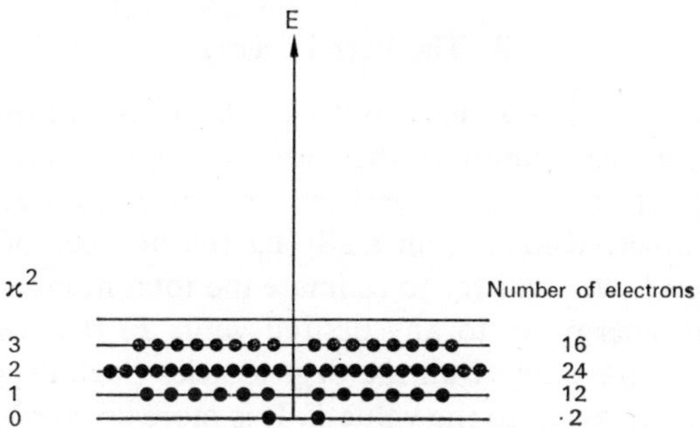

FIG. 15. Free-electron levels and their degeneracies.
Each dot is one electron of either spin.

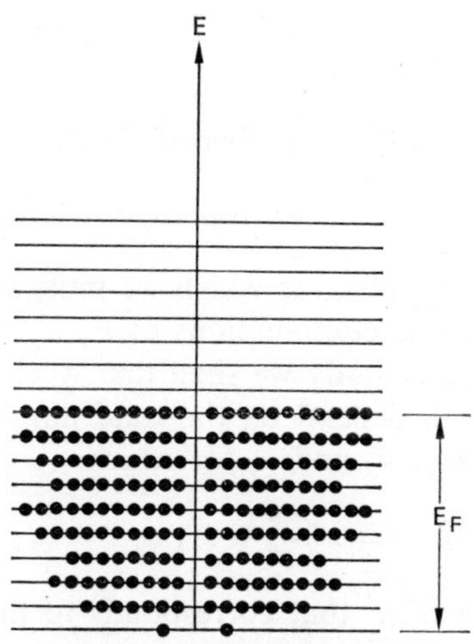

FIG. 16. The Fermi energy.

corresponds to an energy which is characteristic of the given metal, and which is called the *Fermi energy* E_F. We shall see later that this is of the order of 5 eV for a typical metal, whereas the spread of the levels for a classical gas is, as discussed in § 2, about 0·02 eV: this shows how dramatic the effect of the Pauli principle is.

C

8. The Fermi energy

In principle, if we know the number of electrons and the degeneracy of each level, we can compute the Fermi energy. In practice, such a detailed study of the energy levels is impossible on account of their enormous number. Rather than studying the number of electrons in each energy level, we shall try to estimate the total number of electrons which have an energy up to any desired value E. It is clear that this number, $\nu(E)$, is the sum of all the degeneracies (including spin) of all the levels with energy up to the value E. It is more convenient, however, to work with the concentration $\mathcal{N}(E)$ of such electrons:

$$\mathcal{N}(E) \equiv \nu(E)/\text{volume} = \nu(E)/L^3. \tag{32}$$

Since all the electrons in the metal must occupy states below the Fermi energy E_F, we must have

$$\mathcal{N}(E_F) \equiv \mathcal{N} = \text{number of electrons/volume}$$

$$= \text{electron concentration}. \tag{33}$$

We want to estimate $\nu(E)$, for which we must study the degeneracy arising from the possible combinations of κ_x, κ_y, and κ_z (see Table 1). In order to simplify this study we shall first work in two dimensions, in which case $\nu(E)$ will depend on the total number of combinations of all possible values of κ_x and κ_y for which

$$\kappa_x^2 + \kappa_y^2 \leqslant s^2, \tag{34}$$

where s^2 is the value of κ^2 that corresponds to the given value of E,

$$E = h^2 s^2 / 2mL^2 \text{ [cf. (31)]}.$$

If we take for the sake of an example $s^2 = 16$ we can represent graphically the possible states, as shown in Fig. 17, where κ_x and κ_y are given along respective perpendicular axes.

We mark with a dot in the figure all points (i.e. all pairs of values κ_x, κ_y), which are possible, that is which satisfy the condition $\kappa_x^2 + \kappa_y^2 \leqslant 4^2$. As an example, for the point A, $\kappa_x^2 + \kappa_y^2 = 13$. Actually, since we can also take negative values of κ_x and κ_y, the figure should be extended symmetrically into all four quadrants. We notice that all the points that can be used are contained within a circle of radius s (equals 4 in the figure).

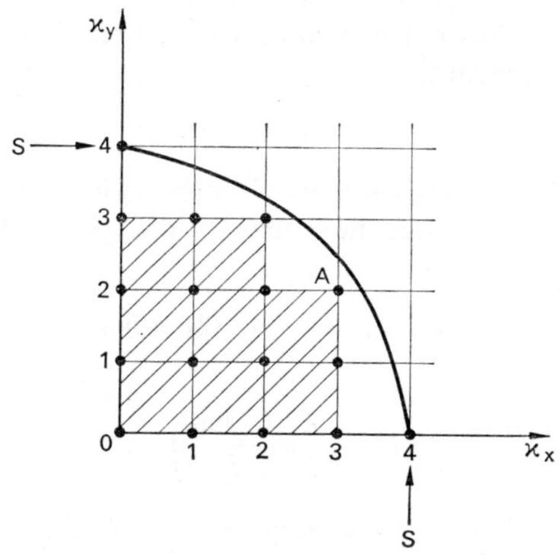

Fig. 17. Grid to count the possible
energy states.

$\nu(E)$ will be twice (because of the spin) the number of such points, which can now be easily counted as follows: there is one point per each square shown in the figure, since each of the four corners of the square is shared by four squares (except for some points near the edge of the circle, but see later). Therefore:

$$\text{number of points} = \text{number of squares} \simeq \text{area of circle}. \quad (35)$$

The last step is justified as follows: the area of each square is unity. Hence, the number of squares is numerically equal to the area shaded in the figure, which, when s is large, is very nearly equal to the area of the circle. (The reader can verify that the area of the circle in Fig. 17 is approximately 50, in units of area equal to the area of a square, and that the number of points is 45. The difference, due to the squares on the boundary, is of course negligible for the large values of s which are required in a real metal, where the total number of electrons is of the order of the Avogadro number.)

The passage from two to three dimensions, i.e. to the proper consideration of κ_x, κ_y, and κ_z, is very simple: the circle of Fig. 17 becomes a sphere and the number of possible points will be given by the volume of the sphere of radius s. This radius, from the relation between it and E given before, is equal to

$$s = L(2E\mathbf{m}/h^2)^{\frac{1}{2}}. \tag{36}$$

The number of states $\nu(E)$ will be given by twice the corresponding number of points (i.e. twice the volume of the sphere), because of the spin. Therefore

$$\nu(E) = \frac{8}{3}\pi s^3 = \frac{8}{3}\pi L^3 (2E\mathbf{m}/h^2)^{3/2}, \tag{37}$$

and from (32)

$$\mathcal{N}(E) = \frac{8}{3}\pi \left(\frac{2E\mathbf{m}}{h^2}\right)^{3/2}. \tag{38}$$

In order to obtain the Fermi energy E_F we take the value of E equal to E_F in (38) and, remembering that $\mathcal{N}(E_F)$ on the left-hand side is the electron concentration $\mathcal{N}$ [from (33)], we obtain

$$E_F = \frac{h^2}{2\mathbf{m}}\left(\frac{3\mathcal{N}}{8\pi}\right)^{2/3}. \tag{39}$$

PROBLEM. Copper is face-centred cubic (4 atoms per unit cell) and its lattice constant a is $3\cdot6$ Å. Assume that each atom contributes one electron to the Fermi gas and hence compute the Fermi energy in electron volts.

$$h = 6\cdot6 \times 10^{-27} \text{ erg sec}, \quad \mathbf{m} = 9\cdot1 \times 10^{-28} \text{ g}, \quad 1 \text{ erg} = 0\cdot63 \times 10^{12} \text{ eV}.$$

9. The density of states

Let us call g'_{E_i} the degeneracy (including spin) of the level E_i. Since we defined (see the beginning of § 8) $\nu(E)$ as the sum of all the degeneracies (including spin) of all the levels up to the value E, we can write

$$v(E) = \sum_i g'_{E_i}, \tag{40}$$

where the sum is over all levels for which $E_i \leqslant E$. Just as we changed $v(E)$ into $\mathcal{N}(E)$ by means of (32), we shall replace g'_{E_i} by the corresponding value per unit volume, i.e. by

$$g_{E_i} \equiv g'_{E_i}/L^3. \tag{41}$$

We observe that, since the separation of the energy levels is very small (of the order of 5 eV divided by the Avogadro number), the energy levels form a quasi-continuous distribution. Therefore, rather than considering the degeneracy of the level E_i as in (41), i.e. the number of states with energy E_i per unit volume of metal, it is more significant to discuss the number of states per unit volume of metal in a small range of energy ΔE_i around E_i. We shall represent this number with the symbol $n'(E_i)$:

$n'(E_i) =$ number of electrons of either spin per unit volume of
           metal in the energy range ΔE_i around E_i. (42)

A yet more significant quantity is obtained by referring the above one to a unit energy range:

$$n(E_i) = \frac{n'(E_i)}{\Delta E_i} = \text{number of electrons of either spin per unit}$$

volume of metal in the energy range ΔE_i around E_i per unit range of energy. (43)

This quantity is called the *density of states* $n(E_i)$ at the energy E_i, and it is very important since it can be compared directly with experimental results, as will be shown in the next section.

We shall now show how to compute the density of states for free electrons. From the definition of $n(E_i)$ it follows that $n(E_i)\,\Delta E_i$, or, in the limit of small quantities, $n(E_i)\,dE_i$, is the number of electrons per unit volume of metal in the energy range dE_i. Therefore $\int_0^E n(E_i)\,dE_i$ is the total number of electrons per unit volume of metal with an energy up to E. This is the quantity which we called $\mathcal{N}(E)$ in § 8 [see (32)].

Therefore

$$\int_0^E n(E_i)\,dE_i = \mathcal{N}(E),\tag{44}$$

from which it follows at once that

$$n(E) = \frac{d\mathcal{N}(E)}{dE}.\tag{45}$$

On differentiating $\mathcal{N}(E)$ as given by (38), we obtain

$$n(E) = \frac{8\pi}{3}\left(\frac{2\mathbf{m}}{h^2}\right)^{3/2}\frac{dE^{3/2}}{dE},$$

that is

$$n(E) = 4\pi\left(\frac{2\mathbf{m}}{h^2}\right)^{3/2}E^{\frac{1}{2}}.\tag{46}$$

This is a parabola as a function of E, which we depict in Fig. 18. Since no levels can be occupied above the Fermi level, the density of states for the occupied levels will be the thick curve of the figure.

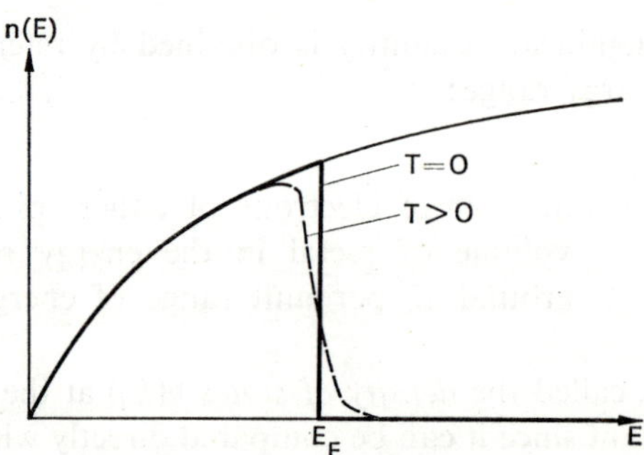

Fig. 18. Density of states curve of a free-electron gas.

Actually, this will be the case only for $T = 0$. At higher temperatures the thermal spread of 0·02 eV will blur a little the Fermi energy edge and we shall have a curve such as the dotted one in the figure, where it can be seen that a small number of electrons have been transferred to states of slightly higher energy.

We shall consider in the next two sections two types of experimental results which can easily be understood by means of the concepts of the free-electron theory of metals that we have developed so far.

10. Soft X-rays

The density-of-states curve of Fig. 18 can be compared directly with the intensities of soft X-ray emission, as we shall now see, thus providing a very powerful check of the previous theory.

Let us consider the aluminium atom, the configuration of which in the ground state is $(1s)^2 (2s)^2 (2p)^6 (3s)^2 3p$. We represent its energy levels in the usual way in Fig. 19a. The outer electrons $(3s)^2 3p$ are delocalized in the metal and take part of the Fermi gas, whereas the inner levels are almost unaltered. The three valence electrons, when delocalized, give place to a sequence of levels such as that depicted in Fig. 16, which we represent in Fig. 19b. If there are n aluminium atoms in the metal, there must be in it $2n$ $1s$ electrons, as indicated in the figure, and similarly for the other levels. The Fermi gas must contain $3n$ electrons up to the Fermi level, as shown. In Fig. 19c we give the density-of-states curve for the Fermi gas (cf. Fig. 18, which is of the same type, except for the orientation of the axes).

Suppose now that the metal is bombarded with cathode rays of enough energy to dislodge one of the electrons in the $2p$ level. One of the electrons of the Fermi gas will fall into the vacant level, as is well known from the theory of atomic spectra, and will emit radiation (X-rays) of frequency $v = \Delta E/h$, where ΔE is the drop in energy experienced by the electron in going from the metal level to the core $2p$ one. Since electrons in any of the levels of the Fermi gas can decay into the $2p$ level, the emitted lines will form a band. Let us now consider the intensities of the various parts of this band. They must be proportional to the number of electrons in the particular level from which the electron is decaying, so that, for instance, the lines a and b in the figure will have very different intensities which reflect the difference in the density-of-states $n(E_a)$ and $n(E_b)$ of the corresponding level. There must be other factors that affect the line intensity but if we assume for the time being (see later) that they are roughly constant throughout the band, the intensities of the emission lines will follow fairly closely the density-of-states curve. In fact since the metal will be at a temperature $T > 0$, we shall have a curve like the dotted one in Fig. 19d (cf. Fig. 18).

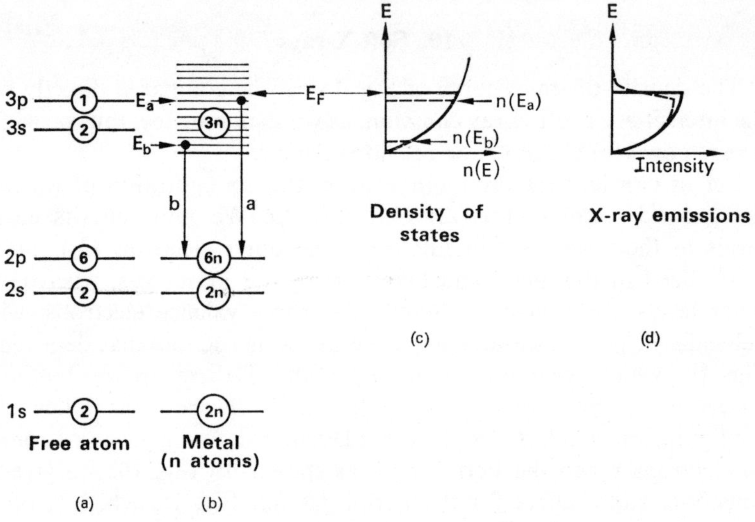

FIG. 19. Origin of the soft X-ray spectrum of aluminium. The figures in circles give the number of electrons in the corresponding states.

We show in Fig. 20 the experimentally observed intensities for aluminium metal, as determined by O'Bryan and Skinner: the similarity with the density-of-states curve of Fig. 18 is striking. In fact, we can also make a quantitative comparison between the value of the Fermi energy which can be estimated from Fig. 20, of about 16 eV, and the one calculated from (39) which is 12 eV, in fair agreement.

A more precise interpretation of this type of X-ray spectrum is nevertheless very difficult. This is so because it is not true that the density of states is the only variable factor throughout the levels of the Fermi gas. The nature of the wave function that corresponds to each state is also relevant and affects the so-called transition probabilities given by quantum mechanics, the variation of which can be even more significant than that of the density of states. Also, the experimental curves tend to tail off at the low-energy end and this blurring affects the accuracy with which the *band width* or Fermi energy can be determined.

Nevertheless, the semi-quantitative interpretation of the soft X-ray emission spectra of metals was one of the early triumphs of the free-electron theory.

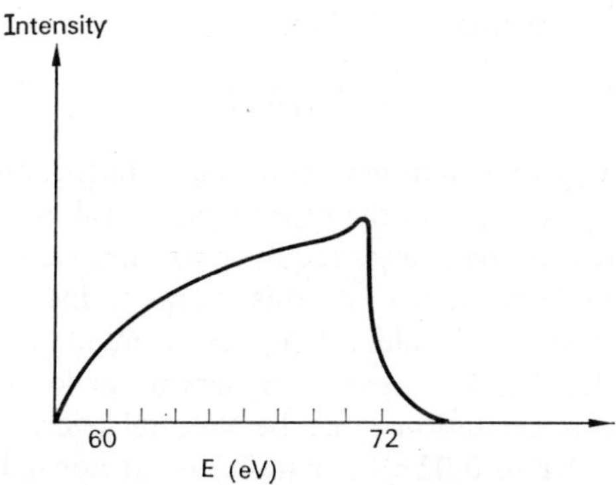

FIG. 20. Intensity of X-ray emission for Al (L_{III}). (After O'Bryan and Skinner, *Phys. Rev.* **45**, 370 (1934).)

It should be appreciated that the band widths to be approximately measured are of the order of 5 eV, as the Fermi energy is. On the other hand, some X-ray transitions, such as those that correspond to jumps between levels in the Fermi sea and the innermost atomic levels (such as $1s$ in Fig. 19), correspond to energy changes of the order of 1000 eV. Clearly, when dealing with such transitions, it is not possible to determine the band width with enough accuracy. The experimental study must therefore concentrate on transitions to high-lying core levels, that give rise to X-rays of low energy, which are therefore called *soft X-rays*. The study of such transitions was pioneered by H. W. B. Skinner.

C*

11. Heat capacities

The study of the *electronic heat capacity* of metals, that is of that part of the heat capacity which can be attributed to the Fermi gas, presented a very hard problem in the early theory of metals before the operation of the Pauli principle was understood, since the calculations were in disagreement with the experimental results by a factor of the order of 100.

Let us first revise the concept of the *molar heat capacity* C_v of a classical gas. We know that $C_v = \partial E/\partial T$, where E is the energy of a mol of gas. This, which is purely kinetic is

$$E = \tfrac{3}{2} k_B N_A T, \tag{47}$$

where N_A = Avogadro number and k_B = Boltzmann's constant. From (47) $C_v = \tfrac{3}{2} k_B N_A$. On the other hand, it follows from the definition of C_v that it is the energy required to increase the temperature of the gas by one degree and to do this means to increase the average kinetic energy of the molecules of it, which entails exciting, by and large, all the molecules of the gas to higher energy levels (see Fig. 21). The source of this excitation must be thermal energy, which as we know is of the order of 0·02 eV per molecule at normal temperatures. We notice in Fig. 21 that all the particles of the gas are accessible for this purpose, that is that all of them can be promoted to higher energy levels by the expenditure of the available quantity of energy (0·02 eV). This, however, is not the case for a Fermi gas: an electron in a level, such as a in the figure, cannot be excited by an energy as low as 0·02 eV, since there is no unoccupied level within this range of energy from it. To excite such an electron, an energy of the order of 0·1 eV is required, which cannot be supplied thermally (remember that since $C_v = \lim_{\Delta T \to 0} \Delta E/\Delta T$, ΔE must be a small quantity). It is only an electron in a level such as b that can be excited into an unoccupied state by a perturbation of the order of 0·02 eV. Such electrons occupy a narrow band, about 0·02 eV in width, just below the Fermi energy. This narrow band is hatched in the figure.

We have now seen that not all the particles of a Fermi gas can take part in the process whereby C_v is measured. Although the gas must contain 1 mol of electrons (since we are dealing with the molar heat capacity), it will appear for this purpose to have a much smaller effective number, namely that of the electrons that occupy levels in the hatched region of the figure. Very roughly, this number will be $(\Delta E/E_F)N_A$.

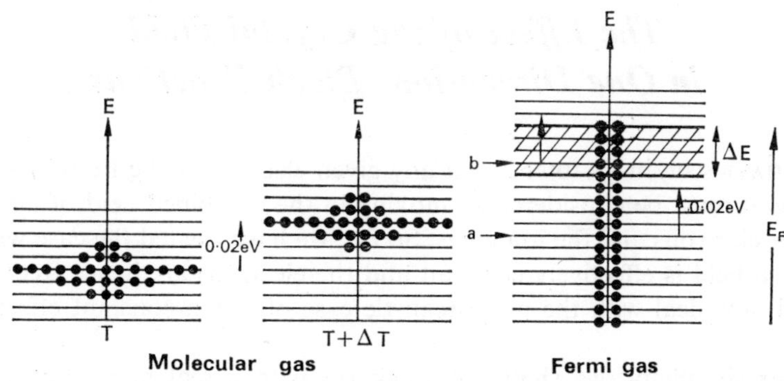

FIG. 21. Thermal excitation of a gas. (The arrow shown represents a jump of 0·02 eV in energy. This is also the width of the region hatched.)

In the Fermi gas, therefore, we must substitute this number for N_A in (47) to obtain the energy E' of that part of the gas which is effective in the process: $E' \simeq \frac{3}{2} k_B N_A (\Delta E/E_F) T$, which leads to a corrected value C'_v of the heat capacity $C'_v \simeq \frac{3}{2} k_B N_A (\Delta E/E_F)$. Since $\Delta E/E_F \simeq 0·02\,\text{eV}/5\,\text{eV} = 1/250$, the corrected heat capacity C'_v of the Fermi gas is smaller than the classical value by a factor of the order of 250, which explains the marked disagreement between classical theory and experiment. This was one of the early triumphs both of the free-electron theory of Sommerfeld and of quantum mechanics.

CHAPTER 3

The Effect of the Crystal Field in One Dimension: Bloch Functions

WE HAVE seen in Chapter 2 that any given electron of the Fermi gas of a metal must be considered as moving in the combined field of all the other electrons and the ion cores. So far, we have treated the case when such a field is effectively constant and the electrons are hence free: we shall now deal with the more realistic case of a non-constant effective field.

For simplicity, we shall not work yet with a real metal but shall rather consider a one-dimensional model, consisting of a *linear chain* of identical atoms repeated at identical intervals of length *a* (*lattice constant*) (see Fig. 22). Consideration of the three-dimensional case will be deferred until Chapter 4.

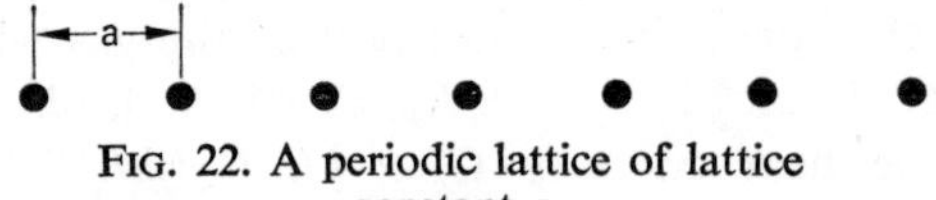

FIG. 22. A periodic lattice of lattice constant *a*.

Clearly, the effective potential field in which the electron moves along the one-dimensional lattice must have exactly the same periodicity as the lattice itself. We shall recognize (see § 3) that this periodicity involves the existence of translational symmetry and we shall discuss in § 1 how and why we want to exploit this symmetry.

1. The use of translational symmetry

We have already seen in § 1.13 that the existence of symmetry in a system is responsible for the fact that some of its energy states are degenerate. This is why symmetry is such an important physical concept.

62

Furthermore, we shall show in this section that the study of symmetry provides an important tool in order to deduce the general form of the solutions of the Schrödinger equation.

1. SYMMETRY OPERATIONS

The symmetry property which we recognized in the example of § **1.13** was the existence of a centre of inversion. If we look again at Fig. 8 (p. 20) we can see that such symmetry exists because two points with abscissa $+d$ and $-d$, say, respectively, are entirely equivalent in the problem in question. (This means that all physical properties in the problem, such as the potential field, must have the same values at these two points.) Likewise, if we consider the infinite chain of which a part is shown in Fig. 22, all points that are at the same distance d of a lattice point are physically equivalent. (Not so, of course, in a finite chain, since some points would be nearer the end of the chain than others.)

The existence of symmetry in such a case is best tested by the concept of *covering operation*. Imagine that a tracing of Fig. 22 has been made (on an infinite strip of transparent paper) and that this tracing is displaced until a lattice point in the tracing overlaps the first lattice point to its right in the original. It can be seen at once that each lattice point overlaps an equivalent lattice point and it is said that the original figure has been *covered*. The operation that brings about the covering in this particular example is called a *translation* by the length a (the lattice constant). (Notice that, since both the lattice and its tracing are infinite, no end lattice points exist that would be left uncovered, as would be the case for a finite lattice.)

In general, symmetry operations will be defined as *covering transformations* of all the points of a space, i.e. as transformations whereby the transformed points cover always equivalent points. Important operations in the linear chain are the translations by a length R (a multiple of the lattice constant a), which transform a point x into $x + R$. These translations will be denoted with the symbol **R**, which will be further discussed below. Another symmetry operation that must be considered in this system is the inversion **i** which transforms x into $-x$. When two operations are performed in succession they sometimes cancel each other and as a net result the system is unchanged. It is convenient to say that the *identity* operation (represented conventionally with the symbol **E**) has been performed.

2. SYMMETRY OPERATORS

Under a symmetry operation the coordinate x of a point in the space will take a new value, x', say. As in § 1.6, it is more convenient to denote the point x' with a symbol that can be made to refer explicitly to the symmetry operation that has brought about this change. Thus, we shall write $x' = \mathbf{G}x$, where the symbol **G** is described as the *operator* that transforms x into x' and can, by suitable choice, be associated uniquely with each symmetry operation. For example, the translation by a length R will be described by an operator **R** such that $\mathbf{R}x = x + R$ and the inversion by an operator **i** such that $\mathbf{i}x = -x$.

It is important to realize that symmetry operations have a dual role: when the operator **G** is applied on a given space, not only the variable x changes into $x' = \mathbf{G}x$, but also any function f of the variable x changes into a new function f', say. It is convenient to designate this function also with the symbol $\mathbf{G}f$ to denote explicitly the operation that has produced this function from the original one f. For example, the inversion along the x axis entails the coordinate transformation $x \rightarrow -x$ and, accordingly, the function $\sin x$ experiences the transformation $\sin x \rightarrow \sin(-x) = -\sin x$, which is written as $\mathbf{i}\sin x = -\sin x$. Of course, the transform of a function is not always as simply related to the original function as in this case. In fact, the sine function is an example of a function with a particularly simple behaviour with respect to the inversion, since, as the function $u(x)$ in Fig. 23a, it is an *odd function*, for which

$$\mathbf{i}\,u(x) = u(-x) = -u(x) = (-1)\,u(x). \tag{1}$$

On the other hand, (b) in Fig. 23 is an *even function* for which

$$\mathbf{i}\,g(x) = (+1)\,g(x). \tag{2}$$

Equations (1) and (2) are of the form of eigenvalue equations (see § 1.6) so that even and odd functions can be considered as eigenfunctions of the inversion operator with eigenvalues $+1$ and -1 respectively.

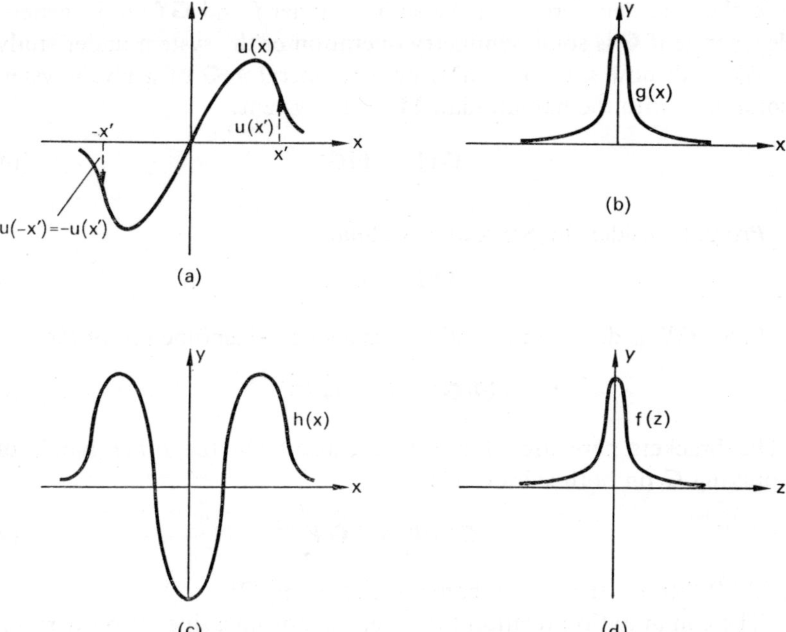

FIG. 23. Transformation of functions under inversion. Even and odd functions. The z axis in (d) is perpendicular to the yx plane in (b).

Likewise, eigenfunctions of the translations can be obtained, which will be done in § 4, and they occupy a central position in the theory of solids under the name of *Bloch functions*.

3. THE EIGENFUNCTIONS OF THE SYMMETRY OPERATORS ARE EIGENFUNCTIONS OF THE HAMILTONIAN

We saw in § 1.13 that a symmetry operation cannot alter the energy of a system and that two functions related by a symmetry operation are degenerate. For example, the two free-electron eigenfunctions given in (1.44) and (1.45), $\varphi_k = \exp(2\pi ikx)$ and $\varphi_{-k} = \exp(-2\pi ikx)$ can now be related by the transformation

$$\mathbf{i}\,\varphi_k = \mathbf{i}\exp(2\pi ikx) = \exp(-2\pi ikx) = \varphi_{-k}, \tag{3}$$

and they are degenerate. In the same manner f and $\mathbf{G}f$ are in general degenerate if $\mathbf{G}$ is some symmetry operation of the system under study.

We shall now show that a symmetry operator $\mathbf{G}$ of a given system commutes with the hamiltonian $\mathbf{H}$ of the system:

$$\mathbf{GH} = \mathbf{HG}. \tag{4}$$

Proof. Consider the Schrödinger equation

$$\mathbf{H}\Psi = E\Psi. \tag{5}$$

Since $\mathbf{G}\Psi$ is degenerate with Ψ it is also an eigenfunction of (5):

$$\mathbf{H}(\mathbf{G}\Psi) = E(\mathbf{G}\Psi). \tag{6}$$

(The brackets here are of no significance.) On the other hand, on applying $\mathbf{G}$ on both sides of (5),

$$\mathbf{GH}\Psi = E\mathbf{G}\Psi, \tag{7}$$

and (4) follows at once on comparing (6) and (7).

The commutation relation (4) is very important: since two operators that commute have common eigenfunctions (§ 1.12), the eigenfunctions of a symmetry operator are automatically eigenfunctions of the corresponding hamiltonian. This allows us to simplify enormously the solution of the Schrödinger equation (5) for the linear chain since, as we shall see later, the eigenfunctions of the translations (Bloch functions) are very easily obtained.

4. *S*-DEGENERACY

So far, we have used the symmetry operators very much as the operators that were introduced in order to represent dynamical variables in quantum mechanics. This is quite right, but there is one important difference between the two types of operators, which we shall now discuss. This concerns the degeneracy of the eigenfunctions. Consider, for instance, the eigenvalue equation (2), $\mathbf{i}\,g(x) = (+1)\,g(x)$, which defines even functions. We must recognize that the number of independent eigenfunctions of $\mathbf{i}$ that satisfy this equation is infinite, since *all* even functions are solutions of it. For example, the function $h(x)$ in Fig. 23c satisfies (2) and is independent of $g(x)$ since, clearly, it is not a multiple of it. In our previous terminology, $g(x)$ and $h(x)$ are degenerate and, because no symmetry operation would ever change one into the other, this degeneracy is accidental. This example is quite typical: for symmetry operations accidental degeneracy of infinite order is the rule rather than the exception. For this reason, it is customary to take such a degeneracy for granted when dealing with symmetry operators and to refer to two functions as degenerate with respect to a symmetry operation if and only if they are related by some symmetry operation of the system under study. In order to avoid confusion the following terminology will be used in this book. Two eigenfunctions of a symmetry operator that correspond to the same eigenvalue of the operator will be called *S-degenerate* if they are related by some symmetry operator of the system under study. Otherwise, they will be called *non-S-degenerate*.† It should be noticed that this terminology must *not* be used except for symmetry operators.

An example of *S*-degeneracy is given in Fig. 23. The functions $g(x)$ and $h(x)$, although belonging to the same eigenvalue $(+1)$ of the inversion operator, are non-*S*-degenerate. On the other hand, if the system under study admits of a rotation by $\frac{1}{2}\pi$ around the y axis of the figure as a symmetry operation, then $g(x)$ and $f(z)$ are degenerate with respect to **i**, since they belong to the same eigenvalue of it $(+1)$ and they are related by a symmetry operation of the system.

It should be clear why we want to sift out the non-*S*-degenerate functions. We are doing all this work in order to identify later on the eigenfunctions of symmetry operators with eigenfunctions of the energy. When the eigenfunctions of symmetry operators are regarded as eigenfunctions of the energy, it is only the *S*-degenerate functions that can become degenerate eigenfunctions of the energy. The non-*S*-degenerate functions, which are unrelated by symmetry, can produce at most accidental degeneracies and must lead in general to non-degenerate eigenfunctions of the energy.

---

† The terminology of *S* and non-*S*-degeneracy is not an established one. It has been introduced in this book purely to help the reader. A full study of symmetry properties requires the tools of group theory in which the functions that we have called *S*-degenerate are said to belong to so-called *irreducible bases*.

5. The importance of continuity

The above considerations are very important in understanding the nature of the process upon which we are now going to embark. This is so because, as we have said, we shall find the eigenfunctions of **H** in terms of those of the translation operators **R** and in doing so we must foresee the following difficulty. Since the free-electron eigenfunctions $\exp(2\pi i k x)$ form a quasi-continuous distribution with respect to k, we can expect a quasi-continuous distribution of eigenfunctions to exist as well, at least for some ranges of k, in the present case. Imagine, however, that for two very close eigenvalues of the translations, say λ_1 and λ_2, we find two eigenfunctions φ_1 and φ_2 which are also very close. (That is, which are almost identical.) From the point of view of the translations alone we could just as well replace φ_2 by any one of the infinite number of non-S-degenerate eigenfunctions that belong to λ_2, but if we did so the continuity of the eigenfunctions would be lost. We must therefore proceed very systematically: we shall start with one eigenfunction of **R** and move away from it to find all others that can be obtained by a continuous (or more properly quasi continuous) variation. In doing this we must be careful to exclude the many non-S-degenerate functions which, although acceptable from the point of view of the eigenvalues of the translations, would spoil the continuity. All this will be a great deal clearer when the work is actually done.

2. S-degeneracy of the eigenfunctions of the translations

We shall now prove that the eigenvalues of the translations **R** of the linear chain are non-S-degenerate, i.e. that given an eigenvalue λ of **R** and an eigenfunction φ that belongs to it, all other eigenfunctions of **R** that belong to λ are such that either they are identical with φ (except for a constant factor) or they are not related to φ by any symmetry operation of the linear chain. In order to do this we must study the commutation properties of the operations of the linear lattice. We shall describe these operations in detail in § 3, but it should nevertheless be fairly clear by now that the only symmetry operations available are the translations and the inversion at the origin. The two following results should be clear from Figs. 24 and 25:

(i) Two translations always commute.

(ii) A translation $\mathbf{R}$ does not commute with the inversion $\mathbf{i}$.

We can now prove the main result of this section. Suppose that φ_k and φ_k' are two S-degenerate eigenfunctions that correspond to the same eigenvalue λ_k of a translation $\mathbf{R}$. (Here k is some numerical label used to designate the eigenvalue.) For S-degeneracy to exist there must

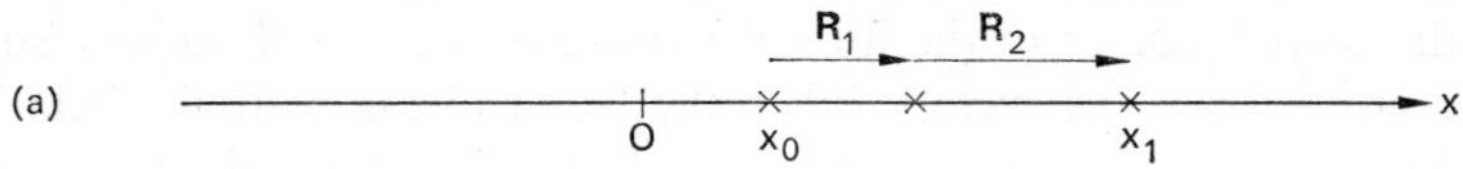
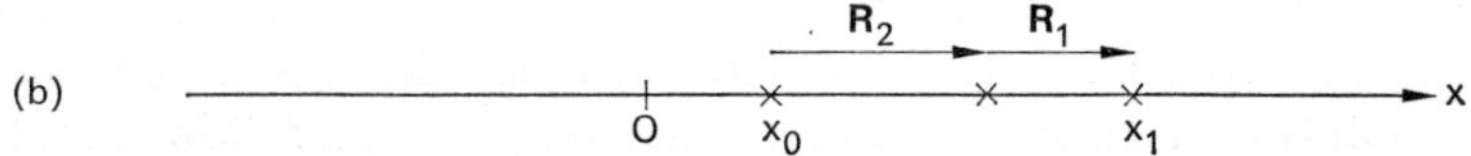

FIG. 24. Two translations always commute. In (a) $\mathbf{R}_1$ is applied first and followed by $\mathbf{R}_2$. In (b) $\mathbf{R}_2$ is applied first and followed by $\mathbf{R}_1$. In both cases the point x_0 is transformed into the same point x_1.

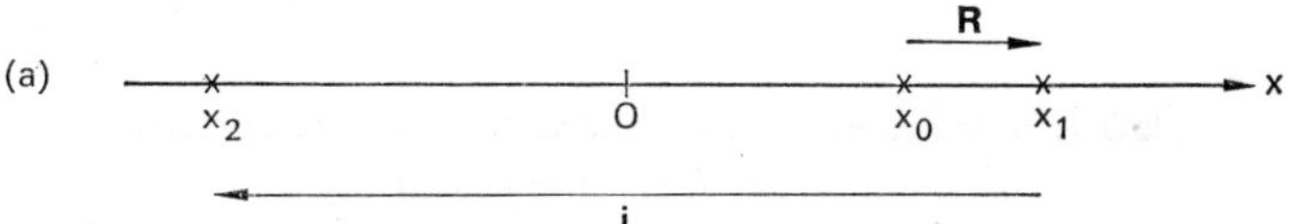
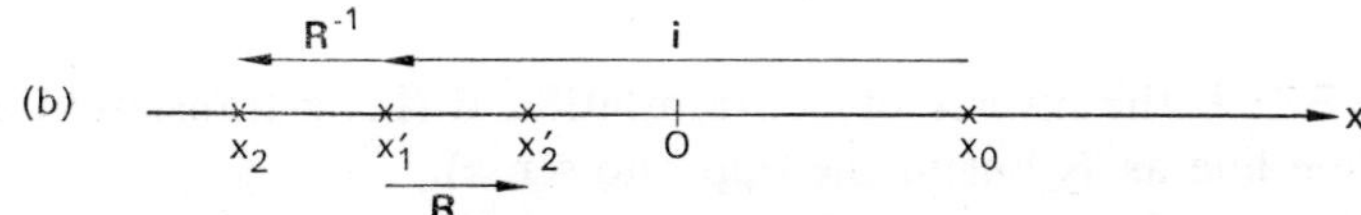

FIG. 25. Translations do not commute with the inversion. In (a) $\mathbf{R}$ is applied first and followed by the inversion. x_0 is transformed into x_2. In (b) the inversion is applied first and followed by $\mathbf{R}$. x_0 is now transformed into x_2', different from x_2.

be some operation that transforms φ_k into φ_k'. This must be either some translation $\mathbf{R}'$, for which $\mathbf{R}'\varphi_k = \varphi_k'$, or the inversion: $\mathbf{i}\varphi_k = \varphi_k'$. The first alternative is not possible: it follows from (i) and the theorem in § 1.12 that all translations have the same eigenfunctions, so that $\mathbf{R}'\varphi_k$

cannot be anything else except a multiple of φ_k. As regards the second alternative, since the inversion and the translations do not commute, they do not have eigenfunctions in common so that the relation $\mathbf{i}\varphi_k = \varphi'_k$ is possible in principle. It is not difficult to see, nevertheless, that this relation is unlikely to be valid. When the potential field is allowed to vanish, the eigenfunctions φ_k of the linear lattice must become, in the limit, the free electron eigenfunctions. (Because the linear lattice hamiltonian becomes in the limit the free-electron hamiltonian.) So, if the relation $\mathbf{i}\,\varphi_k = \varphi'_k$ were valid in the linear chain, a similar relation would be likely to be valid for the free-electron eigenfunctions, whereas we know that the latter obey the entirely different relation (3), $\mathbf{i}\,\varphi_k = \varphi_{-k}$. (The reader can find in § 8 a proof that this relation is in fact valid for the eigenfunctions of the linear lattice.) Since there is no symmetry operation $\mathbf{G}$ of the linear lattice such that $\mathbf{G}\,\varphi_k = \varphi'_k$, it follows that two eigenfunctions of the translations that belong to the same eigenvalue λ_k of a translation cannot be S-degenerate.

This result will be important in fulfilling the programme sketched at the end of last section, as will appear in §§ 4 and 7.

●● 1. *Exercise: Non-commutation of translations*
with the inversion ($\rightarrow$ 8)

Show that

$$\mathbf{iR} = \mathbf{R}^{-1}\mathbf{i},$$

where $\mathbf{R}^{-1}$ is the inverse of the translation $\mathbf{R}$ (i.e. a translation along the same line as $\mathbf{R}$ but in the opposite sense).

Remark. It is interesting to notice that, although we have proved in the text that $\mathbf{iR} \neq \mathbf{Ri}$, the present result provides an easy rule to interchange translations with the inversion: it is enough to change $\mathbf{R}$ by $\mathbf{R}^{-1}$ when doing so.

Method. The operator $\mathbf{iR}$ means "apply $\mathbf{R}$ first and then follow with $\mathbf{i}$" (notice that a sequence of operators is always read from right to left, as in log sin α, where the first operation effected on α is the sine).

Thus

$$i\mathbf{R}x_0 = ix_1 = x_2$$

in Fig. 25a. Try to find an operator $\mathbf{R}'$ such that

$$\mathbf{R}'ix_0 = x_2.$$

Notice from Fig. 25b that $\mathbf{R}' = \mathbf{R}^{-1}$, so that, on equating the left-hand sides of the last two equations,

$$i\mathbf{R}x_0 = \mathbf{R}^{-1}ix_0,$$

whence the result follows. When $\mathbf{R}^{-1}$ is substituted for $\mathbf{R}$ in $i\mathbf{R} = \mathbf{R}^{-1}i$, the equivalent expression $i\mathbf{R}^{-1} = \mathbf{R}i$ is obtained.

3. Symmetry operations of the linear chain: translations and inversion

In order to discuss the translation operations in more detail we shall consider as an example a linear chain of six atoms (Fig. 26). It is important to realize that, despite our insistence so far on the importance of the translations, a linear chain such as the one shown in the figure

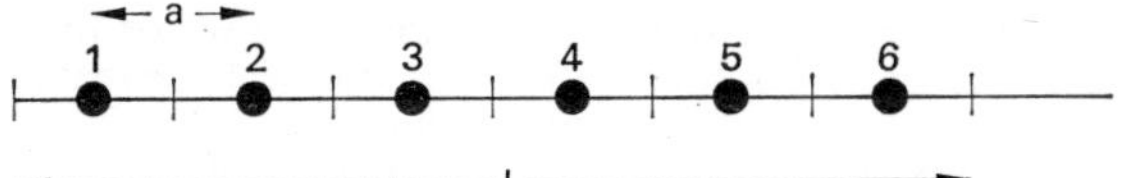

FIG. 26. A linear chain of six atoms. The short vertical lines denote the edges of the unit cell. Notice that the length L of the linear chain has been defined, for convenience, to include as many unit cells in the chain as there are atoms in it.

does not possess translational symmetry. Imagine in fact that we trace Fig. 26 on transparent paper and that we displace the tracing horizontally until atom 1 overlaps atom 2. The tracing does not occupy now a position equivalent to the original one and therefore the displacement effected is not a symmetry operation. Compare, for instance, the similar case of Fig. 27 where, if we rotate the tracing around the centre of the hexagon so that atom 1 overlaps atom 2, the whole tracing overlaps exactly (or covers) the original arrangement.

As discussed in § 1.1, the difficulty in question disappears if the chain is made up of an infinite number of atoms, but this is not convenient since we want to deal with a piece of metal that contains a specified number of electrons rather than an infinite one. However, if we extend

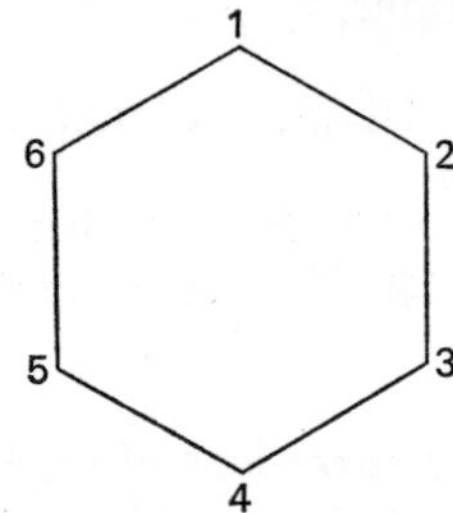

FIG. 27. The linear chain of Fig. 26 bent
so as to preserve periodic symmetry.

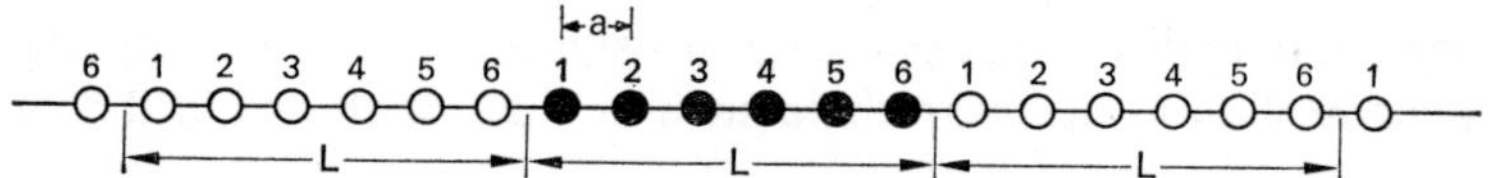

FIG. 28. Born and von Kármán periodic extension of the linear chain (cf. Fig. 13, p. 45). The black atoms represent the real sample of metal of length L.

the finite lattice of Fig. 26 up to infinity *periodically*, as suggested by Born and von Kármán (§ 2.4) and shown in Fig. 28, we can trace the infinite picture and displace it to the right, say, and the tracing occupies an equivalent position, no atom of the tracing being left without another underneath it. This periodic extension of the lattice allows us to have the best of both worlds, i.e. to deal with a finite lattice and yet to use the translational symmetry characteristic of an infinite one. As in § 2.4, the periodicity introduced will result in the existence of some *periodic boundary conditions* for the wave functions.

We shall now want to count the number of possible distinct translations: we can take the tracing of Fig. 28 and leave it undisplaced (the identity operation which will now be called $\mathbf{R}_0$), or displace it by a to the right so that the black atom 1 covers the black atom 2 (a translation by a which will be called $\mathbf{R}_1$), or displace it by $2a$ (translation $\mathbf{R}_2$), and so on. However, when the tracing is translated by $6a$, the black atom 1 sits above the white atom 1 of the first extended period. In fact, the whole of the basic length L sits exactly on top of the length L on its right. We shall agree to identify this translation $\mathbf{R}_6$ with the identity $\mathbf{R}_0$. This is, in fact, required by the Born–von Kármán periodic boundary conditions: since each of the repeated lengths L is identical in every respect to the basic length L, once we displace the tracing of Fig. 27 by $6a$ we have no way whatever to distinguish the new position from the original one corresponding to the operation $\mathbf{R}_0$. (The distinction between black and white atoms in the figure is not, of course, a real one.)

It can be seen from the above that a useful effect of the periodic conditions is to cut down the infinite number of translations of an infinite lattice into a finite number. Thus in our example we have six translations, $\mathbf{R}_0$, $\mathbf{R}_1$, $\mathbf{R}_2$ $\mathbf{R}_3$, $\mathbf{R}_4$, $\mathbf{R}_5$, and $\mathbf{R}_6 \equiv \mathbf{R}_0$. In general, if we have N unit cells in the length $L = Na$, we have N translations which we shall enumerate as follows:

$$\mathbf{R}_1, \mathbf{R}_2, \mathbf{R}_3, \ldots, \mathbf{R}_N = \mathbf{R}_0.$$

In order to describe these operations we shall introduce displacements $R_n = na$. The effect of the translation $\mathbf{R}_n$ on the variable x (measured from a convenient origin along the direction of the linear chain) is to transform x into $x + R_n = x + na$.

The periodic boundary conditions of Born and von Kármán can be justified just as in § 2.4, in particular because they cannot affect the description of a piece of metal that is not too small. In fact, if we bend the linear chain of Fig. 26 into a circle, as suggested in § 2.4, we obtain exactly the regular hexagon of Fig. 27. Here the "translations" that take atom 1 into atom 2, etc., are the successive rotations by $2\pi/6$, $4\pi/6$, $\ldots$, 2π, which correspond exactly to the translations $\mathbf{R}_1$, $\mathbf{R}_2$, $\ldots$, $\mathbf{R}_6 \equiv \mathbf{R}_0$ given before for the linear chain. We have also seen that the periodic boundary conditions agree essentially with the more physical box conditions. Their major merit is that of mathematical convenience, of which we already have some evidence; nevertheless, they are physically artificial and we must not be surprised if their use introduces certain redundancies that will have to be carefully sorted out later on.

Besides the translational symmetry of the linear chain we have to consider also the inversion. Let us take the origin O in Fig. 28 at the midpoint between black atom 1 and the first atom on its left. Clearly, all points equidistant from O are equivalent (§ 1.1) and O is a centre of inversion.

4. The eigenfunctions of the translations

1. BLOCH FUNCTIONS

In a linear chain of N atoms we have N translations $\mathbf{R}_n$ ($n = 1, 2, \ldots,$ N). Since all of these commute, they have the same eigenfunctions; so, in order to find these, it is enough to deal with any one of the $\mathbf{R}_n$ operators. For convenience, we shall denote with $\mathbf{R}$ the operator in question and with R the displacement $R_n = na$ of the x coordinate that corresponds to it.

The eigenvalue equation of R can be written as follows:

$$\mathbf{R}\,\varphi(x) = \lambda\,\varphi(x). \tag{8}$$

On the other hand, since $\mathbf{R}x = x + R$ (§ 3), it follows that†

$$\mathbf{R}\,\varphi(x) = \varphi(x + R), \tag{9}$$

and, on comparing (8) and (9),

$$\varphi(x + R) = \lambda\,\varphi(x). \tag{10}$$

From this expression, we must expect $\varphi(x)$ to be an exponential function of x since, when we add R to the exponent of an exponential, the whole of it is multiplied by $\exp R$ which is a constant (i.e. independent of x). After a little trial and error we can find the most general form of the solution,‡

$$\varphi(x) = \exp(2\pi i k x)\,u(x), \tag{11}$$

---

† The correct derivation of (9) requires a little more attention but the expression that we give, which appears fairly obvious, is sufficient for our purposes.

‡ The details of the exponent in (11) are fairly elastic: we have made a choice that will simplify further work.

where k is some number and $u(x)$ is a function which is periodic with the same periodicity as the lattice [not to be confused with the odd functions of (1)]:

$$u(x) = u(x+R). \tag{12}$$

We shall now verify this solution:

$$\begin{aligned}
\mathbf{R}\,\varphi(x) &= \varphi(x+R) \\
&= \exp\left[2\pi i k(x+R)\right] u(x+R) \\
&= \exp\left(2\pi i k R\right) \exp\left(2\pi i k x\right) u(x) \\
&= \exp\left(2\pi i k R\right) \varphi(x), \tag{13}
\end{aligned}$$

which is of the form (8). The eigenfunctions of the translations (11) are called the *Bloch functions* and the periodic functions $u(x)$ *cell functions* since, on account of their periodicity they are fully defined by their value in any one unit cell of the linear lattice. Their meaning will be discussed in § 6.

It should be noticed that the number k is a label that labels the different eigenfunctions of $\mathbf{R}$ as well as its eigenvalues. In fact, we can rewrite (8) as follows:

$$\mathbf{R}\,\varphi_k(x) = \lambda_k\,\varphi_k(x), \tag{14}$$

where $\varphi_k(x)$ is defined by (11) and where, from (13) and (8),

$$\lambda_k = \exp\left(2\pi i k R\right). \tag{15}$$

We must elaborate further on the details of the eigenfunctions (11). First, we notice that there is nothing in the derivation leading to (13) that precludes $u(x)$ from varying with k. Therefore we must allow this to happen in order to obtain the most general solution possible. Accordingly we shall write, instead of $u(x)$ as before, $u_k(x)$. (This is a function of the variable x that takes different values for different values of the parameter k.) Secondly, we must bear in mind the continuity conditions for the eigenfunctions, which were discussed in § 1.5.

2. CONTINUITY: BANDS

We must remember that the eigenfunctions of the translations are eigenfunctions of the hamiltonian (§ 1.3). Thus an eigenfunction $\varphi_k(x)$ must correspond to an energy eigenvalue that we shall denote as E_k. Take now $\varphi_k(x) = \exp(2\pi i k x)u_k(x)$ and vary k very slightly into k'. It should follow that $\varphi_k(x)$ varies very slightly into $\varphi_{k'}(x)$ and correspondingly E_k into $E_{k'}$. Since $\exp(2\pi i k x)$ varies continuously from k to k', preservation of the continuity will depend on the variation from $u_k(x)$ to $u_{k'}(x)$. The cell functions, however, are very general, since all that we required from them so far was that they satisfy the periodicity condition (12). Therefore, there must be an infinite number of functions $u_{k'}(x)$ all of which are equally suitable to build up the Bloch function $\varphi_{k'}(x)$. We shall select the one that varies continuously from $u_k(x)$. We move along in the same manner from $u_{k'}(x)$ to $u_{k''}(x)$ and select the latter to preserve continuity and so on for the whole of the k axis.

It is now clear that the functions $u_k(x)$ require a further label, so that we shall write them as $u_k^j(x)$. This label is chosen in such a way that when we vary $u_k^j(x)$ into $u_{k'}^j(x)$ continuity is preserved. Likewise, the Bloch functions must now be relabelled:

$$\varphi_k^j(x) = \exp(2\pi i k x)\, u_k^j(x). \tag{16}$$

We must observe that $\varphi_k^j(x)$ and $\varphi_k^{j'}(x)$ belong both to the same eigenvalue λ_k given by (15): when, for a given value of k, we modify the index j, we obtain the infinite number of non-S-degenerate functions which, as we foresaw in § 1, must belong to the same eigenvalue of the translation operator.

When k in $\varphi_k^j(x)$ is varied on keeping the index j constant we obtain a set of Bloch functions that vary continuously one into another with k. Such a set of functions is called a *band*.

We have so far proceeded formally. As a result, it is not at all clear yet what the cell functions are and what the physical meaning of a band is. We shall discuss this in § 6. However, we must realize that the continuity condition from which the concept of a band has arisen is a natural book-keeping device that allows us to classify the multiplicity of non-S-degenerate Bloch functions that belong to the same k: *within a band there is one and only one eigenfunction for each value of k.*

In fact, since the eigenvalues λ_k of the translations are non-S-degenerate (§ 2), it follows that two functions φ_k^j and $\varphi_k^{j'}$ that belong to λ_k must be unrelated by symmetry so that they cannot belong to the same energy eigenvalue. Hence, if φ_k^j and its eigenvalue E_k^j vary continuously from $\varphi_{k+\Delta k}^j$ and $E_{k+\Delta k}^j$, respectively, the same cannot be true for $E_k^{j'}$ (which must differ from E_k^j) with respect to $E_{k+\Delta k}^j$, so that $\varphi_k^{j'}$ cannot belong to the same band as $\varphi_{k+\Delta k}^j$. It follows therefore that φ_k^j and $\varphi_k^{j'}$ cannot belong to the same band either, which confirms the statement made above.

Before we deal with the by now very necessary physical interpretation of the concepts so far introduced we must discuss the quantization of k.

5. Quantization of k

Equations (14) and (15) are valid for all $\mathbf{R}_n$ ($n = 1, 2, \ldots, N$). The eigenvalue λ_k of $\mathbf{R}_N$ must be unity for all k, since $\mathbf{R}_N$ is the identity and cannot alter any of the eigenfunctions φ_k. Therefore, from (15),

$$\lambda_k = \exp(2\pi i k R_N) = \exp(2\pi i k Na) = 1, \tag{17}$$

which requires

$$kNa = \kappa, \quad \kappa = 0, \pm 1, \pm 2, \ldots \tag{18}$$

that is,

$$k = \kappa/Na, \quad \kappa = 0, \pm 1, \pm 2, \ldots \tag{19}$$

Although it appears in a different guise, the condition that λ_k must be unity can readily be seen to be the same periodic boundary condition for the wave function given in (2.18). In fact, since $Na = L$, (19) coincides with the quantization condition (2.20) that was derived from (2.18).

6. Physical considerations

We shall now identify the physical meaning of the quantum number k, of the cell functions $u_k^j(x)$, and of the concept of a band. In order to do this we shall discuss two limiting cases of the linear chain, one where the periodic field is so weak that the solutions must approximate free-electron wave functions, and the other where the interatomic distance a is so large that the atoms of the chain are like free atoms and the wave functions of the linear chain must become atomic eigenfunctions.

1. The Weak-Field Limit

If we allow the potential field of the periodic lattice to diminish until it vanishes, we must expect its wave functions, that is the Bloch functions, to become free-electron eigenfunctions. In order to facilitate this comparison we shall change slightly the notation for the free-electron eigenfunctions given in (2.19) and (2.20):

$$\varphi(x) = \exp(2\pi i k x), \quad k = \kappa/Na, \tag{20}$$

where we have dropped the suffix x and replaced L by its present value Na. We can now compare these functions with the Bloch functions

$$\varphi_k^j(x) = \exp(2\pi i k x) u_k^j(x), \quad k = \kappa/Na. \tag{21}$$

[See (16) and (19).] We notice immediately that the functions (21) go in the limit into the functions (20) if the cell functions $u_k^j(x)$ are allowed to become a constant for very weak fields.

Since, as we have just seen, the Bloch functions become free-electron eigenfunctions in the limiting case of a vanishing potential field, we can now give a physical meaning to the label k which we use in designating them. So far in this chapter, k has been no more than a label or quantum number to designate the eigenfunctions of the translations (and therefore of the energy). We see that in the limit of zero potential it coincides with the wave number of the free-electron wave that then obtains. Therefore k has the limiting meaning of a wave number or, with suitable units, of a momentum (since for free electrons $p = hk$). For this reason, k is sometimes called the *quasi-momentum*, terminology that will be the more appropriate the weaker the potential field of the metal.

2. The Free-Atoms Limit

We now allow the lattice constant a to go to infinity: $a \to \infty$. It follows from the expression for k given in (21) that $k \to 0$ for all κ. Therefore, the exponentials in the Bloch functions (21) become unity for all k and

$$\varphi_k^j(x) \to u_k^j(x) \quad \text{for all } k. \tag{22}$$

In the limit, of course, we must take the value $k = 0$ in the limiting function $u_k^j(x)$, which for simplicity we shall represent with the symbol $u^j(x)$. This means that, as we expand the lattice, all the Bloch functions of a band (which, as we shall see in § 7 are N in number) go in the limit into the single function $u^j(x)$. On the other hand, in the limit of infinite separation between the atoms of the linear chain, the linear-chain eigenfunctions must become atomic eigenfunctions: $u^j(x)$ must coincide with one of the free-atom eigenfunctions. The superscript j designates which one of the infinite free atom eigenfunctions is the limit of the Bloch functions of a band. We can now understand the meaning of the band index j, as well as that of a band itself.

To understand this important point better, let us study the opposite process. We start with a chain of, say, hydrogen atoms that are infinitely apart and assume that all the hydrogen atoms are in exactly the same atomic state ϕ^j, which can be any one of the well known $1s$, $2s$, $2p$, etc., orbitals. At this separation ϕ^j can be identified with the cell function u^j and, as we compress the lattice, ϕ^j is gradually deformed and takes the form of the cell function that is appropriate to the lattice, $u_k^j(x)$. In doing this we shall ultimately generate the N Bloch functions of the jth band. If, instead, we start with all the hydrogen atoms in the state ϕ^r, we shall generate the N Bloch functions of the rth band.

Summarizing: the cell functions $u_k^j(x)$ have the asymptotic meaning of atomic eigenfunctions. When we expand a lattice, the N Bloch functions of a band deform continuously into the same limiting atomic function. The band label indicates this asymptotic behaviour.

The relation between the energies of the Bloch functions of a band and the atomic levels is illustrated in Fig. 29. Two points should be noticed in the figure. The first one concerns the *band width*, i.e. the difference in energy between the top and bottom levels of a band. This quantity has to be obtained by detailed computation, but we can expect certain atomic levels to be hardly deformed at all in the metal so that the corresponding band would be very *narrow*. This will be the case for inner-shell orbitals (such as $1s$ and $2s$ for sodium) which are so compact that the normal interatomic distances are very large with respect to them. Therefore, even in a normal metal these orbitals are in a situation which is very nearly that of infinite separation and the corresponding band collapses into a single energy level or very nearly does so. In general, the greater the interaction between the orbitals in neighbouring atoms, the greater the band width will be. In sodium, for instance, we expect the *3d band* (i.e. the band arising asymptotically from the $3d$ level) to be narrower than the $3s$ band, since the latter orbitals are more extended in the free atom.

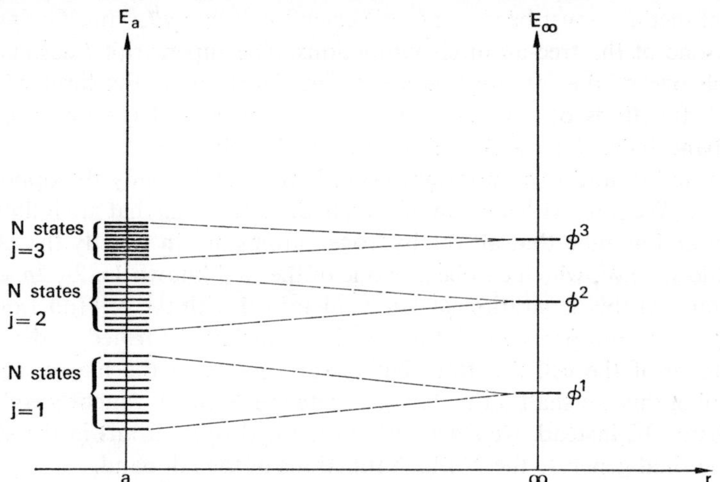

FIG. 29. Broadening of atomic levels into bands. *r* indicates the interatomic distance. On the left, the levels of the first three bands are represented at the metal interatomic separation *a*. When this is increased up to infinity, all the levels of each band collapse into the same level of the free atom. Cf. Fig. 97 (p. 221) where it is shown how the broadening described here is calculated.

The second point we want to make about Fig. 29 is this. In principle, the shaded areas which indicate the energy levels could overlap. For instance, we could have a situation such as that shown in Fig. 30, where the hatched area denotes the domain of energy values where the first and second band overlap. We shall see later, however (§ 14), that such overlaps are not normally allowed in the linear chain, although they can appear in three-dimensional crystals.

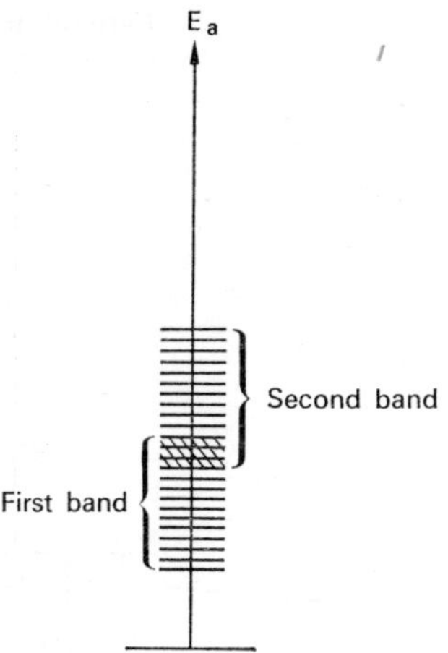

FIG. 30. Band overlap. The hatched
region is one for which there are energy
levels both of the first and of the second
band. This is a *band overlap*, not allowed
in the linear chain with a weak periodic
potential but allowed in three-dimensional metals.

3. ENERGY GAPS

It is reasonable to expect that there must be a one-to-one correlation
between the energy levels obtained in the free-electron limit and those
that correspond to the Bloch eigenfunctions of the lattice field, although
how this is so will not be clear until § 13. We represent this correlation
in Fig. 31, where the free-electron levels are represented as we have
done in Chapter 2 and, for clarity, we have taken $N = 6$. As compared
with the free-electron case, the effect of the crystal potential can be
see to be that of creating *forbidden energy gaps* (as long, of course, as
overlap is not allowed, as mentioned before). These forbidden energy
gaps can be regarded as vestiges of the gaps between the atomic levels:
their existence has very important physical consequences, which we
shall discuss later (§§ 16, 18).

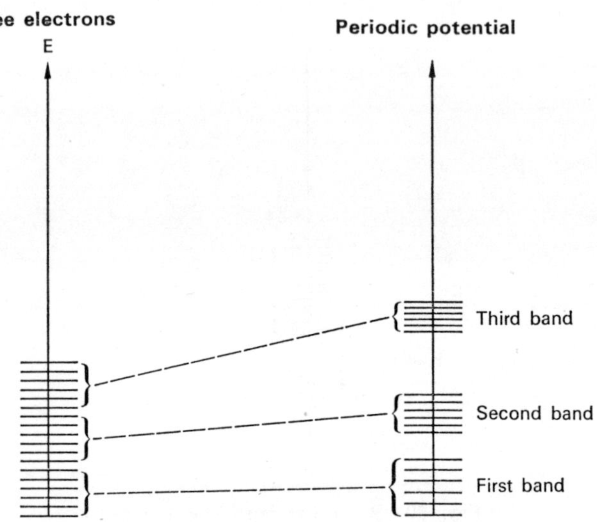

FIG. 31. One-to-one correspondence between the levels in the free-electron case and those for a periodic potential.

7. The number of eigenfunctions

We shall verify in this section that, as advanced in § 6, there are N Bloch functions in each band, that is exactly as many as there are unit cells in the lattice.

1. THE NUMBER OF DIFFERENT EIGENVALUES λ_k IS N

We shall first count the number of eigenvalues λ_k given by (15). Since in (15) $R = na$ (n an integer),

$$\lambda_k = \exp(2\pi i k n a). \tag{23}$$

We shall prove that

$$k' = k + m a^{-1}, \tag{24}$$

where m is an arbitrary integer, leads to exactly the same eigenvalue as k itself. That is

$$\lambda_{k'} = \lambda_k. \tag{25}$$

Substitute k' for k in (23):

$$\lambda_{k'} = \exp\left[2\pi i(k+ma^{-1})na\right] = \lambda_k \exp\left(2\pi imn\right) = \lambda_k,$$

since mn is an integer.

Since, from (19), $\kappa = kNa$, k' corresponds to a value of κ, κ' say, such that $\kappa' = k'Na = kNa+mN = \kappa+mN$. We have therefore proved that two values of κ that differ by a multiple of N lead to the same eigenvalue λ_k. Hence, the only values of k that lead to distinct eigenvalues λ_k are

$$k = \frac{\kappa a^{-1}}{N}, \quad \kappa = 1, 2, \ldots, N, \tag{26}$$

since all other values of κ differ from those given in (26) by a multiple of N. We have thus shown that there are at most N different eigenvalues of $\mathbf{R}_n$, since there are no more than N values of k that will lead to different eigenvalues. It should be clear that the N eigenvalues of $\mathbf{R}_1$,

$$\lambda_k = \exp\left(2\pi ika\right) \quad \text{with} \quad k = \kappa a^{-1}/N \quad \text{and} \quad \kappa = 1, 2, \ldots, N, \tag{27}$$

[from (23) with $n = 1$] are all different, whereas at the other end of the range for n all the eigenvalues of $\mathbf{R}_N$ are equal to unity. (As expected, since $\mathbf{R}_N$ is the identity.)

2. Periodicity in k

It follows from (24) that the quantum number k has a periodic structure: we can add to it a multiple of a^{-1} without changing the eigenvalue denoted by k. a^{-1} is therefore a sort of lattice constant as far as k is concerned. For convenience we shall denote this basic period a^{-1} with the symbol K and one of its multiples ma^{-1} with $K_m = mK$. The periodicity in k expressed by (25) and (24) can now be written more compactly:

$$\lambda_k = \lambda_{k+K_m}. \tag{28}$$

This periodicity in k will be discussed further in § 10.

D

3. The number of Bloch functions is N

The important result so far is that $\mathbf{R}_1$ possesses N different eigenvalues λ_k for N possible values of k. We wanted to count the number of eigenfunctions of the translations: since the eigenfunctions of $\mathbf{R}_1$ are also eigenfunctions of $\mathbf{R}_n$ for all n, it is just enough to count the eigenfunctions of $\mathbf{R}_1$.

It follows from § 2 that the N different eigenvalues λ_k of $\mathbf{R}_1$ are non-S-degenerate. This means that, given a function φ_k^j that belongs to λ_k, any other function that belongs to λ_k must be either identical with φ_k^j (except perhaps for a constant factor) or entirely unrelated to it. Since $\varphi_{k+K_m}^j$ belongs to λ_k [from (28)], it must either be identical to φ_k^j or, being unrelated to it by symmetry, it must be one of the functions $\varphi_k^{j'}$ that belong to k but to a different band j'. However, $\varphi_{k+K_m}^j$ denotes a function of the jth band, so that the natural choice between the two alternatives given is to take $\varphi_{k+K_m}^j = \varphi_k^j$. It follows therefore that the N different eigenfunctions

$$\varphi_k^j(x) = \exp\,(2\pi i k x)\, u_k^j(x) \text{ for } k = \kappa a^{-1}/N \text{ and } \kappa = 1, 2, \ldots, N, \quad (29)$$

are all the possible eigenfunctions that we must consider in a given band.

The reader is now in a position to appreciate the importance of our discussion of S-degeneracy. Unless we had gone through the rather subtle arguments leading to it, we would not have been able to identify properly $\varphi_{k+K_m}^j$ with φ_k^j. If S-degeneracy in k had been permitted, for instance, both these functions might have been S-degenerate and therefore not necessarily identical (or independent).

It should be noticed that the relationship between φ_k^j and $\varphi_{k+K_m}^j$ can be established in a different way. One of the alternatives that we mentioned before was that $\varphi_{k+K_m}^j$ could be identified with $\varphi_k^{j'}$. This possibility, of course, involves a change of notation, since so far we have accepted that φ_k^j for all k is a function of the jth band. This is why we rejected this possibility in the first instance. It is nevertheless possible and, indeed, sometimes convenient, as we shall see in § 13, to introduce a change of notation so as to define $\varphi_{k+K_m}^j$ as the function $\varphi_k^{j'}$. This involves no contradiction, if we accept that the superscript j denotes a given band only as long as we keep to the basic set of N functions listed in (29) and that when k is incremented by K_m we jump into a different band. We therefore recognize that we have two choices:

$$\varphi^j_{k+K_m} \equiv \varphi^j_k , \tag{30}$$

$$\varphi^j_{k+K_m} \equiv \varphi^{j\,\prime}_k . \tag{31}$$

Nevertheless, as we have seen, (30) is the natural choice to make and we shall consider it the fundamental one. The alternative provided by (31) appears at this stage unorthodox and somewhat obscure. In fact, some details of it require further development and it is only when we do this in § 13 that we shall be able to make sense of this alternative. Unless explicit statement to the contrary, (30) will always be assumed to be used in what follows.

8. The effect of the inversion

A plausibility argument was given in § 2 to show that the inversion transforms the Bloch function φ_k that belongs to the eigenvalue λ_k into the Bloch function φ_{-k} that belongs to λ_{-k}:

$$\mathbf{i}\,\varphi^j_k(x) = \varphi^j_{-k}(x). \tag{32}$$

●● 1. THE INVERSION CHANGES k INTO $-k$: PROOF

We must first recognize that, from (14) and (15), the label k of an eigenfunction $\varphi_k(x)$ merely indicates the nature of the eigenvalue in the expression

$$\mathbf{R}\,\varphi^j_k(x) = \exp\,(2\pi i k R)\,\varphi^j_k(x). \tag{33}$$

Equation (33) can be used for the operator $\mathbf{R}^{-1}$ (see the Exercise in § 2.1) for which the translation vector is $-R$. Thus

$$\mathbf{R}^{-1}\,\varphi^j_k(x) = \exp\,(-2\pi i k R)\,\varphi^j_k(x).$$

Operate on both sides of this equation with the inversion operator $\mathbf{i}$, remembering that it does not affect the exponential since this is not a function of x:

$$\mathbf{i}\,\mathbf{R}^{-1}\,\varphi^j_k(x) = \exp\,(-2\pi i k R)\,\mathbf{i}\,\varphi^j_k(x).$$

From the exercise in § **2**.1 we know that $\mathbf{i}\mathbf{R}^{-1} = \mathbf{R}\mathbf{i}$, whence

$$\mathbf{R}[\mathbf{i}\,\varphi_k^j(x)] = \exp(-2\pi i k R)\,[\mathbf{i}\,\varphi_k^j(x)].$$

(The square brackets have no particular meaning: they are merely inserted for convenience.) On comparing this equation with (33) it is clear that the function in square brackets satisfies the eigenvalue equation of $\mathbf{R}$ with the eigenvalue $\exp(-2\pi i k R)$. Since the function that pertains to this eigenvalue must be labelled $\varphi_{-k}^j(x)$, (32) follows.

2. The role of the Bloch functions

We can now comment on our work so far. We showed in § 1.3 that, since symmetry operators commute with the hamiltonian, the eigenfunctions of the latter can be obtained by forming the eigenfunctions of symmetry operators. Not all symmetry operators can be used at once, since not all of them commute (translations, for instance, do not commute with the inversion, as shown in § 2), and it is only those that commute which will have common eigenfunctions. All translations commute and we have formed their common eigenfunctions in order to obtain from them the eigenfunctions of the energy. However, the reader may have wondered why we did not start instead by building eigenfunctions of the inversion. Equation (32) is the answer to this. This equation is an instance of a general result which can be proved by group theory: the set of eigenfunctions of the translations (Bloch functions) is sufficient in order to express the results of all other symmetry operations of the lattice. This means that the extra symmetry operations of the lattice merely transform the Bloch functions amongst themselves [as in (32)] or into combinations of themselves, without introducing any new eigenfunctions. (See § 4.7 for a discussion of this result for three-dimensional lattices.) This result places the eigenfunctions of the translations in a privileged position with respect to those of other symmetry operators.

9. Degeneracy of the Bloch functions

The reader must notice that the prefix S was not used in the heading. This is so because we do not want to discuss now the degeneracy of a symmetry operator but rather that of the energy.

Since the translations commute with the hamiltonian (see § 1.3), their eigenfunctions, i.e. the Bloch functions (29), are eigenfunctions of the energy:

$$\mathbf{H}\,\varphi_k^j(x) = E_k^j\,\varphi_k^j(x), \tag{34}$$

where quite naturally we label the energy eigenvalues with the same labels as those of the eigenfunctions. It is convenient, however, to introduce a small change of notation. In fact, since $k = \kappa a^{-1}/N$ with $\kappa = 1, 2, \ldots, N$, it follows that for N of the order of the Avogadro number the discrete values of k are very finely spaced and can be regarded as a quasi continuous variable. It is therefore more convenient to write $E^j(k)$ for E_k^j, to parallel the notation normally used to denote dependence on a continuous variable.

The question we want to discuss is the degeneracy of these energy eigenvalues, and in doing so we shall exclude accidental degeneracies from our consideration since nothing much can be said about them.

We first notice that, because the eigenvalues of the translations are not S-degenerate, there is only one function φ_k^j for a given k and j, and that two functions φ_k^j and $\varphi_k^{j'}$ can never be degenerate eigenfunctions of the energy because no symmetry operation will ever transform one into the other. On the other hand, although φ_k^j and $\varphi_{k'}^j$ cannot be S-degenerate eigenfunctions of the translations (since they must belong to different eigenvalues of them) they can be degenerate eigenfunctions of the energy as long as there is some symmetry operation that relates them. This cannot be a translation since we know that translations leave the k number of a Bloch function invariant (since the Bloch functions are their eigenfunctions). The only other symmetry operation that can introduce degeneracy is the inversion $\mathbf{i}$, and it follows at once from (32) that φ_k^j and φ_{-k}^j are degenerate. In summary:

$$E^j(k) \neq E^{j'}(k), \tag{35}$$

$$E^j(k) = E^j(-k). \tag{36}$$

At last we begin to reap the fruit of our labours: as we can see, the formal work carried out so far in this section allows us to obtain important results about the energy eigenvalues without having to solve the Schrödinger equation. In fact, we now possess all the required tools and we shall see in the next few sections how much more information can be obtained in this way.

10. Periodicity of the energy

It follows from (28) and the discussion in § 7.3 that two values of k that differ by K_m must be considered as indistinguishable:

$$k \equiv k + K_m, \quad \text{with} \quad K_m = mK = ma^{-1}. \tag{37}$$

We have seen, in fact, that the eigenvalues that correspond to k and $k + K_m$ are the same and that the eigenfunctions that belong to them must be taken as identical. Clearly, the corresponding energy eigenvalues must also be the same:

$$E(k) = E(k + K_m). \tag{38}$$

We can see that the value $K = a^{-1}$ of k, which evidently plays an important part in the theory, gives the basic range of values of k, i.e. the range of k which contains the N values that define the N different eigenvalues λ_k. In fact, from (27),

$$k = \frac{a^{-1}}{N}, \frac{2a^{-1}}{N}, \ldots, a^{-1}, \tag{39}$$

so that for large N the N values of k in (39) are very approximately in the range

$$0 \leqslant k \leqslant a^{-1}, \quad \text{i.e.} \quad 0 \leqslant k \leqslant K, \tag{40}$$

and all the values outside this range are equivalent to those within on account of (37).

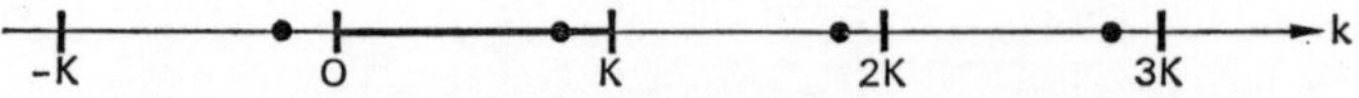

FIG. 32. Periodicity in k. All the values of k marked with a dot are related by (37) and must be considered identical. The unit cell given in (40) is represented with a thick line.

It follows that if we plot the values of k along an axis (k axis), this presents a periodic structure with period $K = a^{-1}$. The length K gives the unit cell in the periodic lattice of k values (Fig. 32). Because of this periodic structure, which parallels that of the crystal lattice, the k axis is often referred to as the k space. It should be noticed that whereas the real x space has period a, the k space has a period $K = a^{-1}$. This is a significant relation, which will appear again in § 4.7.

11. Change of interval

The unit cell in k space which we show in Fig. 32 is not the most convenient to use. We know that, owing to its periodicity, it is enough to plot the energy in the first unit cell in k space. On the other hand, from (36), $E(k) = E(-k)$. In order to display this symmetry it would be convenient to have negative as well as positive k values in the first unit cell in which the energy is plotted. This, of course, is not the case for the cell of Fig. 32, all k values of which are positive in the interval from 0 to K. Fortunately, the periodicity in k allows us to take a new unit cell, centred round the origin, as shown in Fig. 33. This cells goes from $-\tfrac{1}{2}K$ to $\tfrac{1}{2}K$: its half from 0 to $\tfrac{1}{2}K$ was contained in the original cell, whereas its negative half, from $-\tfrac{1}{2}K$ to 0 can readily be seen by periodicity to be identical with the remaining half, from $\tfrac{1}{2}K$ to K, of the original cell.

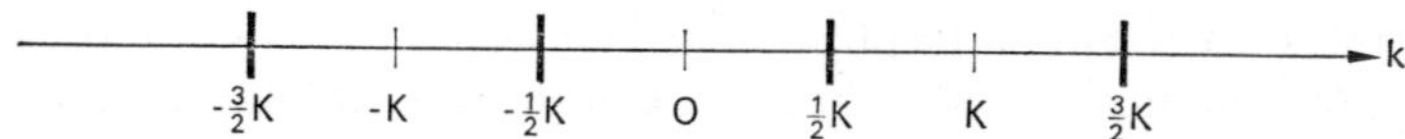

FIG. 33. Unit cells in k. The edges of the new unit cells are marked with thick lines and those of the unit cells of Fig. 32 with thin lines.

The values of $k = a^{-1}/N, 2a^{-1}/N, \ldots, a^{-1}$ [cf. (39)] in the original first unit cell, go over into the following sequence in the new one:

$$k = -\tfrac{1}{2}a^{-1}, \left(-\tfrac{1}{2}a^{-1} + \frac{a^{-1}}{N}\right), \ldots, -\frac{a^{-1}}{N}, 0, \frac{a^{-1}}{N}, \ldots, \left(\tfrac{1}{2}a^{-1} - \frac{a^{-1}}{N}\right). \tag{41}$$

Since a^{-1}/N can normally be neglected, we shall take the first unit cell to be symmetrical, unless good reasons to the contrary exist:

$$k = -\tfrac{1}{2}a^{-1}, \left(-\tfrac{1}{2}a^{-1} + \frac{a^{-1}}{N}\right), \ldots, \tfrac{1}{2}a^{-1}. \tag{42}$$

It should be stressed that the only necessary part of k space is the first unit cell shown in Fig. 33 which contains all the N relevant values of k. All values outside this cell, such as k_i', can be considered to be identical, by periodicity, to some value of k, k_i, say, within the first cell. This means that the Bloch function for k_i' is the same as that for k_i and that $E(k_i')$ $= E(k_i)$. Strictly speaking, therefore, the whole of k space outside the first unit cell is redundant. It is nevertheless convenient to preserve the whole of the k axis because we obviously need it in order to describe free electrons, and if we wish to correlate in detail the results for the periodic lattice with those from the free-electron model, it will clearly be necessary to give some physical meaning to the formally redundant parts of k space. This will be done in § 13.

12. Further properties of the $E(k)$ curve

For most purposes, as we have said, $E(k)$ can be treated as a function of a continuous variable and we shall do so in this section. Further, we shall assume $E(k)$ and its first derivative $dE(k)/dk$ to be continuous within a band. The first assumption is in direct correspondence with the continuity requirements used in § 4.2 to define a band. The second is plausible on account of the physical meaning of dE/dk as a group velocity [see (1.65) and notice that we make use of the physical meaning of k as a momentum, as discussed in § 6.1].

So far, we have the conditions

$$E(k) = E(-k), \tag{43}$$

$$E(k) = E(k+K) \tag{44}$$

It follows from (43) that, in Fig. 34, $E(1) = E(2)$ whence it can readily be seen on taking $\delta \to 0$ that, in order to satisfy the continuity of the derivative, $dE/dk = 0$ at $k = 0$. On applying the periodicity condition (44) this result can immediately be extended as follows

$$dE/dk = 0 \quad \text{at} \quad k = 0, \pm K, \pm 2K, \ldots \tag{45}$$

We now consider the value of this derivative at the edge of the cell. From (43), $E(3) = E(4)$, and from (44), $E(3) = E(5)$. Therefore, $E(4) = E(5)$, and in the same manner as before, $dE/dk = 0$ for $k = \frac{1}{2}K$, a result which can first be periodically extended by (44) and then combined with (45) to give

$$dE/dk = 0 \quad \text{at} \quad k = 0, \pm\tfrac{1}{2}K, \pm K, \pm\tfrac{3}{2}K, \ldots \tag{46}$$

This means that dE/dk vanishes at the centres and edges of all unit cells in k space, that is at all the values of k marked in Fig. 33.

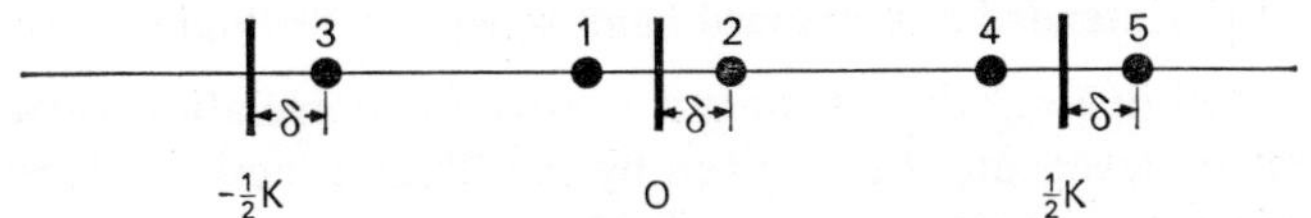

FIG. 34. Values of k in the first unit cell related by symmetry.

The conditions (43), (44), and (46) determine the main features of the $E(k)$ curve. In fact, they require it to be of the general form shown in Fig. 35, with the possible alternative of a similar curve in which the positions of the maxima and minima are interchanged.

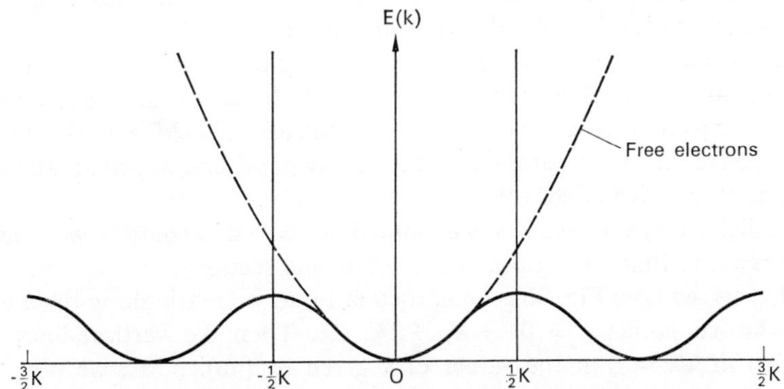

FIG. 35. Energy as a function of k in the periodic lattice. $E(k)$ is drawn within the first unit cell $-\frac{1}{2}K$ to $\frac{1}{2}K$ so that dE/dk vanishes at the centre and edges of it. Outside the first unit cell $E(k)$ is repeated by periodicity. The curve drawn with a broken line gives the $E(k)$ curve for free electrons. All curves, of course, are purely qualitative.

D*

We can compare this energy curve with that for a free electron, the energy of which is $E(k) = p^2/2\mathbf{m} = h^2k^2/2\mathbf{m}$, so that the $E(k)$ curve is a parabola as shown in Fig. 35. It might seem at first that we have lost a very substantial part of the parabola, in fact all of it outside the first cell in k space. However, we must remember that the $E(k)$ curve given for the periodic lattice corresponds to one band only. We must expect that when other bands are introduced a greater degree of correlation with the free-electron parabola will be obtained.

13. Extended and reduced band schemes: Brillouin zones

We stated in § 6.3 that there is a one-to-one correlation between the free-electron levels and those given by the Bloch functions. How this is the case is not quite clear yet. In Fig. 35, in fact, we were able to correlate only one part of the free-electron parabola with the $E(k)$ curve for the Bloch functions. Moreover, the reader may have observed that in § 6.1, when showing how the Bloch functions (21) go over into the free-electron eigenfunctions (20) in the vanishing field limit, we kept discreetly quiet about the fact that, whereas for the free electrons in (20) k runs from $-\infty$ to ∞, for the Bloch functions in (21) k runs from $-\frac{1}{2}K$ to $\frac{1}{2}K$, since, as it turned up in § 11, all values of k outside this range are essentially redundant for the periodic lattice.

We shall now tidy up these loose ends by considering in detail how the solutions of the linear chain correlate with those for the free electrons. In order to do this we shall assume that the crystal field is weak enough for the free-electron parabola to be a fairly good first approximation to the solution. We effect this correlation in Fig. 36, which describes the so-called *extended band scheme* (sometimes called *extended zone scheme* for reasons that will appear later on in this section).

Let us see how Fig. 36 is constructed. First, we mark along the k axis the lattice points $k = 0$, $\pm K$, $\pm 2K$, etc. Then the vertical lines for which $dE/dk = 0$ at the values of k given by (46). Next, we plot the free-electron parabola, and we are ready to try a good guess at the shape of the weak field $E(k)$ curves for the various bands since, the parabola being a good first approximation, we must hug it as much as we can. We start at $k = 0$ and we follow the parabola as far as possible, but when we get near the edges of the first unit cell ($k = \pm\frac{1}{2}K$) we must turn down, to ensure that $dE/dk = 0$ at them. We now have the $E(k)$ curve in the first cell, and we repeat it by periodicity as in § 12.

We must remember that although we have plotted the $E(k)$ curve as a continuous line in the first cell, the true situation is slightly different. There are only N meaningful points equally spaced in each unit cell of the k axis. Thus, there are only N physically significant values of k in the first cell and therefore also N significant values of the energy along the $E(k)$ curve in the first cell, each point, of course, being vertically above one of the permitted k points. The corresponding values of the energy can be read on the vertical scale, and we represent them more explicitly

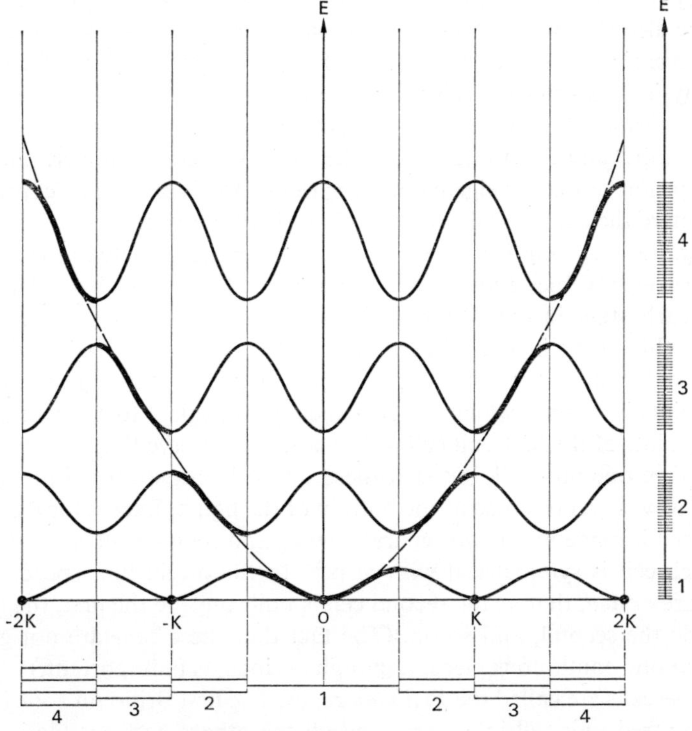

FIG. 36. Extended scheme. Sections of the k axis that belong to the same Brillouin zone are marked with the same number of horizontal lines.

on an auxiliary energy scale to the right of the figure. They are the energy eigenvalues corresponding to the first band. (Notice that although we have N states, that is N eigenfunctions, we have fewer eigenvalues, since the states are doubly degenerate, $\varphi_k^j(x)$ and $\varphi_{-k}^j(x)$ corresponding always to the same value of the energy.)

In order to obtain the second band, we follow the parabola from $\frac{1}{2}K$ to K and in the corresponding negative interval from $-K$ to $-\frac{1}{2}K$. Notice that, again, we draw the curve so as to ensure that the derivative vanishes at the required points. The thick line shown represents $E(k)$ for the second band and we complete it by periodicity. (Notice that in the region of k space which we have used to plot the second band, we have exactly N values of k, because its total length is equal to that of a unit cell. Notice also that we need not worry that the domain of k space used is not the basic one, $-\frac{1}{2}K \leqslant k \leqslant \frac{1}{2}K$, which contains all the values of k which are physically significant. It is clear that all the values of k used are equivalent by periodicity to values in that domain, that is, in the first unit cell.)

The third and fourth bands are drawn in the same manner, and the distribution of energy eigenvalues for the bands is exactly as expected. (Compare the energy scale on the right of the figure with that on the left of Fig. 29.) At long last, the one-to-one correlation between the free-electron eigenvalues and those in the periodic lattice which we depicted at an early stage in Fig. 31 is explained.

The thick lines in Fig. 36 represent the $E(k)$ curves for all the bands as a single-valued function of k, except that there are discontinuities at the points in k space which are periodically equivalent to the centre and edge points of the first unit cell in k space. We observe that each band is plotted in one unit cell, but that these unit cells are disposed in an unusual way, as represented at the bottom of the figure. Except for the first, each cell is made up of two separate pieces and they are all *centred*, that is, each cell is symmetrical with respect to the origin in k space. Also they are *nested*, that is, the second cell is built outside the first, the third outside the second, and so on. (The fact that these cells are not given each in one continuous piece, although curious, is unimportant.)

These cells are called the *Brillouin zones*. The first Brillouin zone is the first centred unit cell in k space and all the others nest outside it. The edges of the Brillouin zones are obtained by the following simple rule, which should be clear from the picture. Consider all lattice points in k

space, i.e. the black circles along the k axis of the figure. Start at the first point ($k = 0$) and bisect the distances from it to its first neighbour lattice points ($k = K$ and $k = -K$). These are the edges of the first zone. Then bisect the distances from the origin to its *second* neighbours ($k = 2K$ and $k = -2K$) and obtain the edges of the *second* zone, and so on. Since all the Brillouin zones are unit cells and therefore of the same length, each of them covers exactly the same length of the k axis and contains exactly the same number N of quantized values of k, where N is the number of unit cells in the corresponding crystal under study.

In the extended band (or zone) scheme that we are describing the energy is given as a single-valued function of k, the first band being plotted in the first zone, the second band in the second zone, and so on. The edges of the Brillouin zones are the points at which $dE/dk = 0$ and at which discontinuities of the energy appear.

When the whole set of wavy curves given in Fig. 36 is considered, it is clear that we have introduced redundancies, since the only physically significant parts of the curves are the thick lines shown. The rest of the curves correspond to states that are not physically distinct from the previous ones. These redundancies (see § 3) are the result of the artificial boundary conditions which we have used but, as we have seen, they do not affect in the least the physical interpretation of the results. In fact, they can be usefully exploited, as we shall see in § 16.

A very useful scheme to represent the $E(k)$ curves is the *reduced band (or zone) scheme*, in which all the bands are represented in the first Brillouin zone, so that the energy is given as a multivalued function of k. We show in Fig. 37 how the $E(k)$ curves of Fig. 36 can be represented in the reduced scheme: it is enough to shift the corresponding parts of the curves back to the first zone by translations by suitable multiples of K, which are indicated in the figure.

It should be observed that the more fundamental of the two schemes is the reduced one since we use in it the only part of k space that is really physically significant (the first Brillouin zone). Also, we do not need to get into details of the higher Brillouin zones which in three-dimensional crystals are quite involved. In practice, when detailed calculations of the bands are made, the results are always plotted in the reduced scheme. On the other hand, for metals which are nearly free-electron like, the extended scheme provides a way of obtaining a simple first approximation with a minimum of calculation, or none at all. Since this qualitative approach was much used in the early days of metal theory, when detailed calculations were beyond the then existing facilities the extended scheme was by far the then most commonly used.

Now that the work has been done we can go back and think about it. The whole trick that we have used depends on the existence of the two alternatives (30) and (31) of § 7.3. If we had known the answer (or at any rate, its general shape) we would have proceeded as follows. Firstly, we would have drawn the bands in the reduced scheme, as in the central part of Fig. 37. This corresponds to using (30) from which condition we

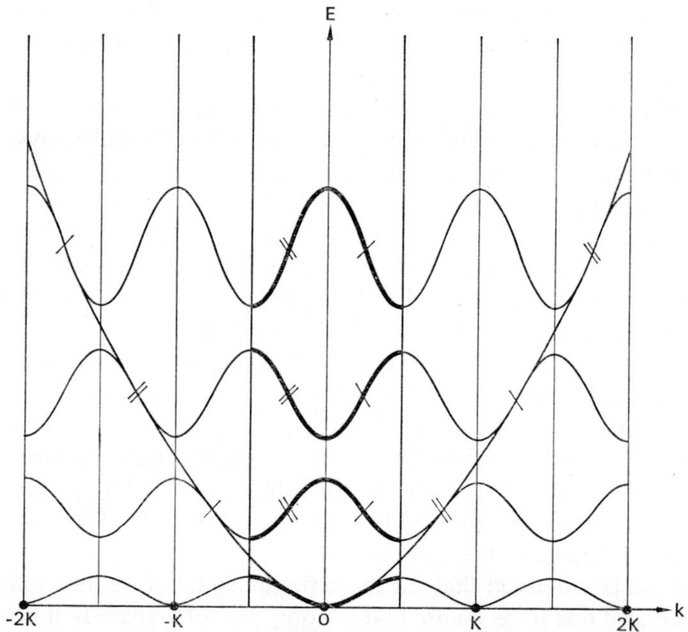

Fig. 37. Reduced scheme. Properly, the only part of the figure that should be given in this scheme is that drawn in thick lines. The rest is given here to show the correlation with the extended scheme of Fig. 36. Each section of the $E(k)$ curve that hugs the parabola of that figure has been translated into the first zone. Sections so related are marked with the same number of strokes.

recognize that nothing outside the first cell is strictly necessary. Secondly, we would have puzzled about how to correlate the resulting figure with the free-electron results. With a bit of luck we would then have recognized that the alternative (31) could be exploited in order to achieve this. When we stated this alternative it did not appear to make much sense. In fact, as we actually stated it, in the form $\varphi^j_{k+K_m} \equiv \varphi^{j'}_k$, it was rather useless since j' was allowed to take any value, so that the right-hand side of the expression was undetermined. At that time, we lacked a rule to fix j'. We can now give a convenient one in order to regain the extended scheme. It is enough to decide that when $K_m = 0$ (i.e. $m = 0$) $j' = 1$ (which means that we plot the first band in the first cell), when $K_m = K$ we take $j' = 2$ (which means that we plot the second band in the second cell), and so on. The rule is simple since it amounts to rewriting condition (31) in the form $\varphi^1_{k+K_m} = \varphi^{1+m}_k$. Thus the reduced and extended zone schemes are two alternative ways to sort out the redundancies which, as advanced in § 3, result from the periodic boundary conditions.

14. Band crossings

The existence of forbidden energy gaps, which was advanced in § 6.3, was confirmed in Fig. 36. There are two ways in which these energy gaps can be removed. The first is through the existence of crossings or points of contact between two bands, and the second through some severe distortions of the bands from the parabolic forms characteristic of weak periodic potentials.

The possibility of crossings will be discussed first. Consider the first and second bands in the reduced scheme of Fig. 37: it can be seen that three relative positions are possible for them, as displayed in Fig. 38a–c. As soon as the bands touch or cross the energy gap disappears, as is illustrated on the left-hand side of each picture. Neither of these situations, however, is allowed in the linear chain. In fact, consider the value k at which the bands 1 and 2 cross in Fig. 38c. At this point the Bloch functions φ^1_k and φ^2_k of the first and second bands respectively are degenerate, which is forbidden by (35) of § 9. (b) can be ruled out in the same manner. (See also Problem 4 in § 24.)

The prohibition of crossings or points of contact is valid for the linear chain with any potential, however strong. Nevertheless, the bands shown in Fig. 38a–c are not too distorted forms of free electron bands. If severe distortion of the bands from their free-electron forms are permitted, the band gaps can be closed as shown in Fig. 38d. Such extreme distortions, however, are most unlikely to occur, as confirmed by experience with fairly strong metal fields. Therefore, it can be generally accepted that the non-crossing rule for the linear chain entails the existence of energy gaps.

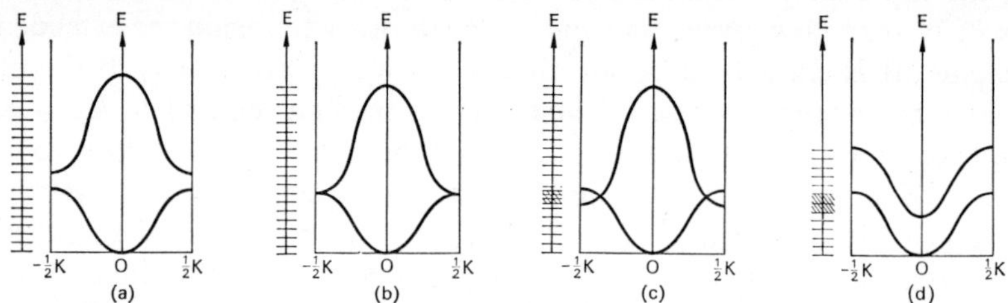

FIG. 38. Band crossings and overlaps. The hatched regions on the left-hand sides of (c) and (d) indicate the overlap between the first and second bands (cf. Fig. 30).

It should be clearly understood, nevertheless, that the non-crossing rule itself is strictly valid only for a one-dimensional lattice: it cannot be guaranteed to hold even for a chain of atoms where the charge cloud around the nuclei is three-dimensional, although the nuclei themselves form a one-dimensional array. The existence of energy gaps at the edges of the Brillouin zone is not, therefore, a necessary condition in three dimensions. The reason for this is that there are now many more symmetries available that can introduce the degeneracies required by the band crossings.

15. Filling in of the energy states

We shall consider as an example linear chains made up of N atoms of sodium, magnesium, and aluminium respectively. The basic properties of these atoms that we want to use is that, since we assume that the inner shells are undistorted in the metal, they contribute one, two, and three electrons per atom respectively to the Fermi gas. Let us assume for the sake of the example that the energy bands are the same in all three cases (also the lattice constant a, which would otherwise affect the size of the first Brillouin zone). We plot the first two bands in Fig. 39.

We know that the number of states in a band (i.e. the number of k values in a Brillouin zone) is N, the number of unit cells in the metal lattice: a full band will contain one state per atom or, with spin, two. (This is no longer the case if there is more than one atom in the unit cell; see § 21.)

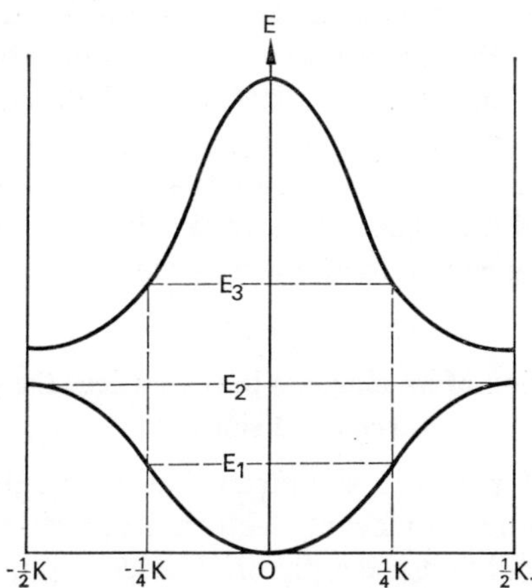

FIG. 39. Filling in of the bands.

These states will be filled in with the available electrons in such a way that the states with lowest energy will be occupied first. Therefore, if we have one electron per atom, as in sodium, we start filling in the states from the centre of the Brillouin zone outwards until the number $\frac{1}{2}N$ of k values (i.e. N states with spin) is occupied. Since the density of possible k values is uniform throughout the Brillouin zone this number of values of k is found in exactly one-half the length of the Brillouin zone. This means that when we have one electron per atom as in sodium we shall move away outwards from $k = 0$ until the values of $k = \pm \frac{1}{4}K$ are reached. We can find at once the energy of the highest occupied level, which we call E_1 in the figure: this will be the Fermi energy that corresponds to one electron per atom. If we have two, as in magnesium, we must fill in the whole of the first band and the Fermi energy is E_2. When we have three electrons as in aluminium, the whole of the first band must be filled first and then we must start filling in the second band. This must be done from the edge of the Brillouin zone inwards since it is in this direction that the energy of the levels increases. We have to accommodate $\frac{1}{2}N$ electrons in the third band so that we must cover exactly one-half the length of the Brillouin zone, from $\frac{1}{2}K$ to $\frac{1}{4}K$ and $-\frac{1}{2}K$ to $-\frac{1}{4}K$. The Fermi energy is E_3.

It should now be clear how the bands are filled in for any electron concentration. Also, that in certain cases bands may be completely full. This is a very important property since, as we shall see in § 18, if the top band is full the crystal is an insulator. (The reader must not draw, from the examples given here, the conclusion that magnesium is an insulator: there are substantial differences that appear in three dimensions, as has already been mentioned in § 14 and will be seen in the next chapter.)

16. Propagation of an electron in the lattice: the periodically repeated scheme

The existence of periodic symmetry has very fundamental consequences from the point of view of electron propagation, which we shall now discuss. We shall first use a pictorial description, in which we shall identify k with the momentum (see § 6.1), although we know that this is correct only for very weak periodic fields.

Whereas the energy of a free electron always increases for increasing positive k, this is no longer the case for an electron in a periodic lattice (Fig. 40). In the region $\frac{1}{2}K \leqslant k \leqslant K$, the larger k is, that is the larger the momentum the smaller is the energy of the bound electron. If we compare with the free-electron curve, we see that such a behaviour is impossible for it except when the momentum is negative. In fact, we know that the whole region $\frac{1}{2}K \leqslant k \leqslant \frac{3}{2}K$ is not really physically significant and has to be folded back on to the first zone, when the range $\frac{1}{2}K \leqslant k \leqslant K$ becomes negative. In this way, a value of k that follows very closely after $\frac{1}{2}K$, such as k' in the figure must be identified with k''.

Consider now an electron propagating along the lattice with a small positive momentum near 0. If we increase its momentum, for instance by applying an electric field along the chain in the positive direction, the state of the electron will be represented by a k value that moves gradually to the right in Fig. 40. When the momentum has been increased until k reaches the edge $\frac{1}{2}K$ of the first Brillouin zone, a curious situation arises. A small increase of the momentum from $\frac{1}{2}K$ brings k up to the value k'. This, however, is indistinguishable from k''. Therefore, when the momentum increases beyond $\frac{1}{2}K$ we must re-start at its equivalent point $-\frac{1}{2}K$ and move to the right from it: i.e. the change from $\frac{1}{2}K$ to k' should be represented as that from $-\frac{1}{2}K$ to k''. We represent the way in which the

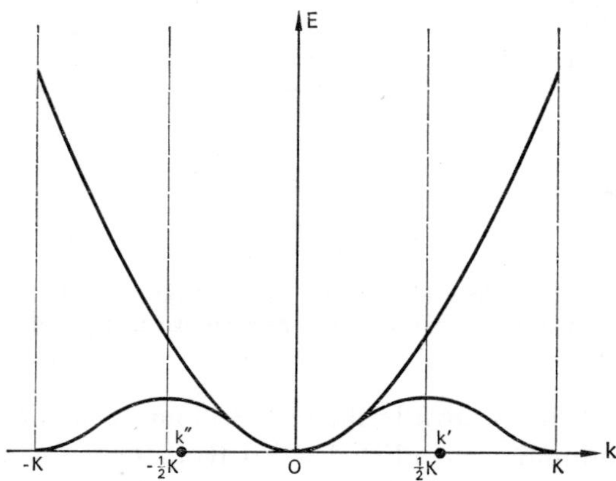

FIG. 40. Periodicity in k.

electron states change by means of the arrow in Fig. 41, where (a) and (b) are entirely equivalent descriptions, except of course that in (a) we keep to the first cell in k space only. It is clear, by considering Fig. 41a that when the electron state reaches the right-hand edge of the first Brillouin zone it is suddenly reflected back, since the states $\frac{1}{2}K$ and $-\frac{1}{2}K$ in (b) are physically indistinguishable. This means that, in this process, the momentum first increases until it reaches the edge of the Brillouin zone, when it suddenly changes from a large positive to a large negative value. This discontinuity, although surprising, is not unusual; it is exactly what happens to a ball reflected back by a vertical wall. This means that when the momentum of the electron reaches the edge of the Brillouin zone, the electron is reflected back by the lattice.

As the electron moves through the crystal, as in the previous examples, its state changes and the point that represents this state in k space moves as represented by the arrows in the lower part of Fig. 41. This path followed by the electron states in k space is called an *orbit* and the reader must beware not to confuse it with the path of the electron in real space.

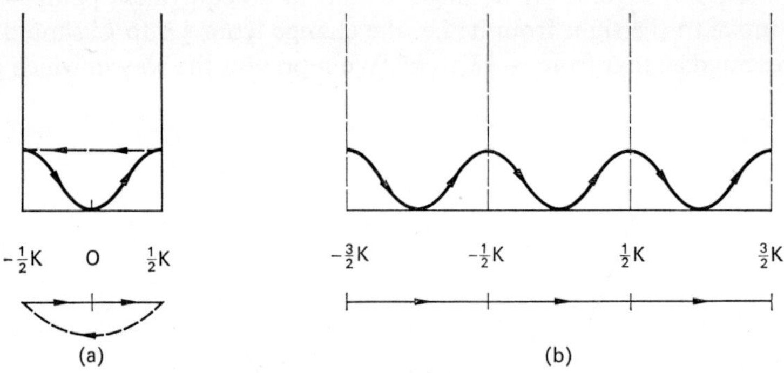

Fig. 41. Reflections in k space. The electron orbits in k space are given in the copies of the k axes shown at the bottom of each figure.

When the orbits are shown in the first Brillouin zone only, they often present discontinuities, as in Fig. 41a. To avoid these discontinuities it is convenient to make use of the redundant, periodically repeated parts of k space, as we do in (b). This figure, which should be compared with Fig. 35 (p. 91) is obtained by repeating periodically the first unit cell (first Brillouin zone) in k space. Such a method for plotting states is called the *periodically repeated scheme* and it is particularly advantageous in two- or three-dimensional cases where, if the orbits are plotted in the first cell only, they can have a very complicated shape on account of the large number of jumps that they show.

In order to discuss in more detail the process of the electron propagation through the lattice, we shall consider the group velocity v_g [see (1.65)]

$$v_g = \frac{1}{h}\frac{dE}{dk}, \tag{47}$$

which is the velocity one must think about when considering the electron as a particle moving in space. If we now look again at Fig. 41b we see that on moving from the origin to the right, dE/dk (and hence the group velocity) increases first to a maximum, from which it decreases to nought at the edge of the Brillouin zone, after which it becomes negative. (The same pattern can of course be followed in Fig. 41a.) We know in fact from (46) that dE/dk must always vanish at the edge of the Brillouin zone, which stresses the physical importance of this edge: when the electron state reaches the edge of the Brillouin zone the velocity of the electron must vanish, which means that the electron wave cannot be propagated by the lattice. Since this state must be followed by another in which the velocity changes sign with respect to the original velocity, the electron wave must be reflected by the crystal, as we had argued before.

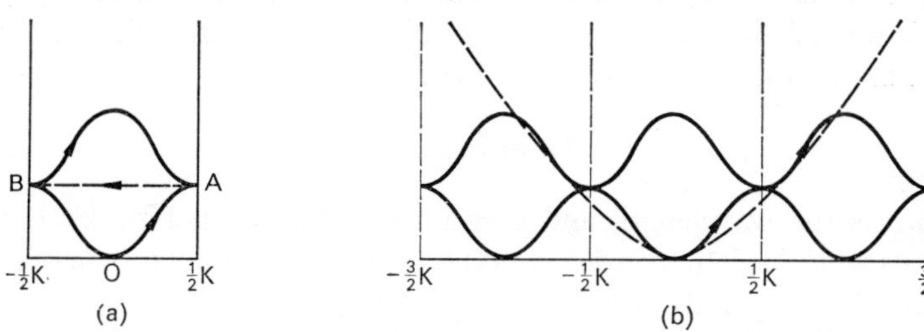

FIG. 42. The reflections in k space lose meaning if there are no energy gaps.

It should be noticed that the reflection of an electron wave at the edge of the Brillouin zone depends on the existence of an energy gap. We see, for instance, in Fig. 42a that although the electron state is reflected from A to B this is no longer significant since the electron is not forced to go down in energy as before. In fact, it can go on increasing its energy along the path indicated by the arrow: this means that dE/dk stays positive, that the velocity is positive, and that the particle is always moving in the same direction. Of course, without the energy gap the electron behaves very much like a free one, as it is quite clear when the same picture is drawn in the repeated scheme, when it is seen that the path closely follows the free-electron parabola. (We know, of course, that in a strictly one-dimensional lattice we cannot have two bands touching as in the example just discussed. The above observations, nevertheless, will be useful in other cases.)

17. Bragg reflections

We have just found that if we have an electron propagating along a linear lattice of lattice constant a, the electron wave is reflected when the momentum of the electron reaches the value $\frac{1}{2}K$ (for the edge of the first Brillouin zone) or, more generally, $\frac{1}{2}nK$, where n is an integer (for the edges of higher order Brillouin zones). Since, from § 7.2, $K = a^{-1}$, the value of the momentum at which reflections occur is

$$k = \tfrac{1}{2}na^{-1}. \tag{48}$$

On the other hand, one knows from elementary physical theory that there is a condition for a wave to be reflected by a periodic structure. This is the well known *Bragg reflection condition*

$$2a \sin \theta = n\lambda, \tag{49}$$

where λ is the wavelength and θ and a are defined in Fig. 43. If our previous work is right, condition (48) should be a particular case of (49). In fact, if $\theta = \frac{1}{2}\pi$, the wave in Fig. 43 is propagating along a one-dimensional lattice of lattice constant a. With this value of θ the Bragg condition (49) gives

$$2a = n\lambda = n/k, \tag{50}$$

from which (48) follows at once.

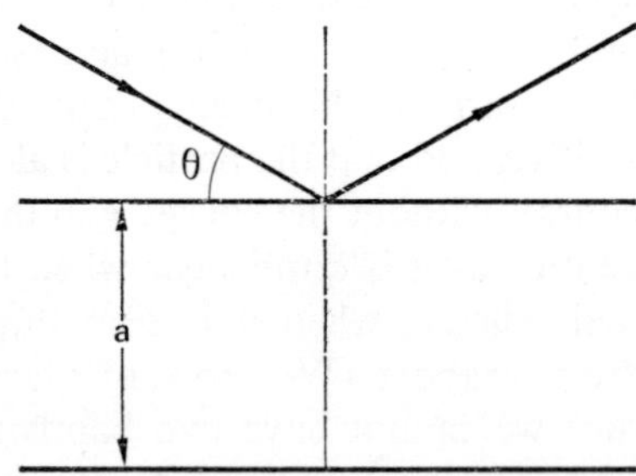

FIG. 43. Bragg reflection.

18. Conductors and insulators

Let us first consider a free-electron metal. We represent in Fig. 44a the $E(k)$ curve for such a case as well as the discrete energy levels corresponding to each k, in an exaggerated scale. Each level corresponds to two degenerate states with values $+k$ and $-k$ respectively and each of these states can accommodate two electrons as shown in the figure: the two electrons on the right have positive momentum hk and those on the left negative momentum.

In the absence of an electric field there are just as many electrons moving to the right as moving to the left, and hence there is no net flow of current. If there is an electric field, say to the right of the figure, its effect is to excite to higher energy levels the electrons near the Fermi energy that are moving to the right. This creates vacancies on the right-hand side of the bottom of the band, which are then filled in by some of the electrons that are moving to the left. The net result (Fig. 44b) is an excess of electrons that move to the right over those that move to the left, whereby a net flow of charge to the right is created.

Exactly the same situation arises in the periodic lattice except that in this case it may not be possible to find empty states above the Fermi level. In Fig. 45a this is not so: the first Brillouin zone is partially filled and

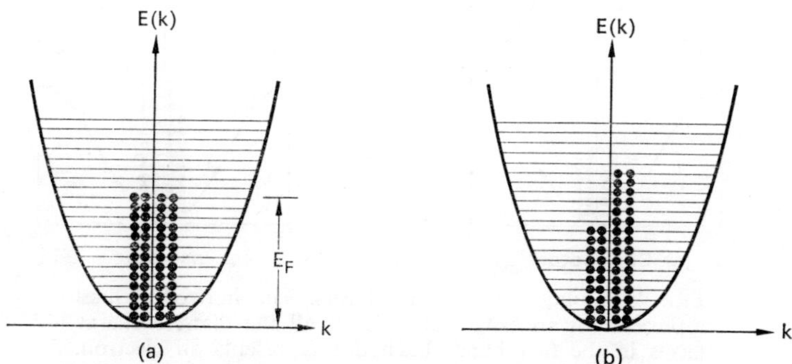

FIG. 44. Distribution of electrons in a free-electron metal. (a) No field. (b) Electric field present. E_F is the Fermi energy. Each dot represents an electron.

there are plenty of available states above the Fermi level: we therefore have a *conductor*. In (b), on the other hand, the only levels above the Fermi one belong to the second band and are therefore separated by an energy gap. In *insulators* this energy gap is so large that it can be surmounted only by electrons excited by fields much larger than the normal ones (when this happens it is said that the *electric breakdown* of the insulator has been produced). On the other hand, if the gap is very small it can be bridged by some excited electrons as in *semi-conductors*.

We have seen in § 15 that for atoms with an even number of external electrons, such as magnesium and calcium, we must expect full bands. It must be clearly understood, nevertheless, that it does not follow that such materials must be insulators: in three dimensions, even without band crossings, the energy gaps may be closed by band overlaps as we shall see in § **4.15**.

On the other hand, we can say in general that a full Brillouin zone will lead to an insulator (or semi-conductor if the next higher band

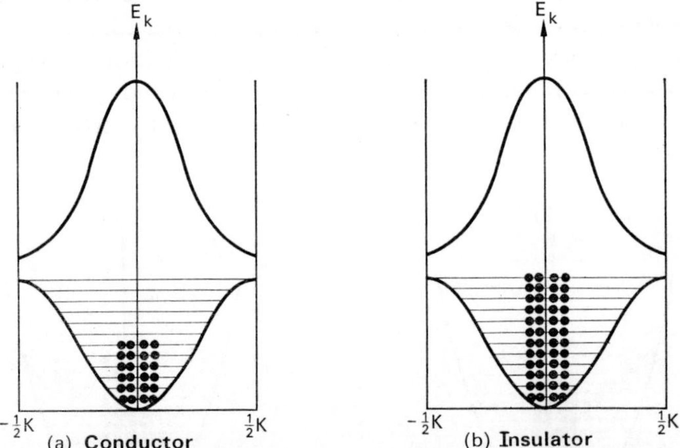

Fig. 45. Conductors and insulators. The horizontal lines represent (in an exaggerated scale) all the discrete energy levels in the first band. Each dot represents an electron. (a) represents a material in which there are not enough electrons in the Fermi sea to fill in the first band. (b) represents a material in which there is exactly the right number of electrons to fill in the first band.

possesses levels very close in energy to those at the top of the occupied band). A partly full Brillouin zone will lead to metallic conduction.

19. Effective mass. Holes

We know that the velocity that must be assigned to an electron is the group velocity $v = (dE/dk)/h$ [see (1.65) from which we now drop the suffix g]. We shall try to compute the electron acceleration a:

$$a = \frac{dv}{dt} = \frac{1}{h}\frac{d}{dt}\left(\frac{dE}{dk}\right) = \frac{1}{h}\frac{d}{dk}\left(\frac{dE}{dk}\right)\frac{dk}{dt}. \tag{51}$$

We shall obtain dk/dt by the following plausible argument. We know that for a classical force f,

$$f = \mathbf{m}a = \mathbf{m}\frac{dv}{dt} = \frac{d}{dt}(\mathbf{m}v) = \frac{dp}{dt}. \tag{52}$$

Since for the electrons in the lattice $p = hk$, it is plausible to write $f = h\,dk/dt$ for the force that acts on them. We substitute the corresponding value of dk/dt in (51) and we find

$$a = \frac{1}{h^2}\frac{d^2E}{dk^2}f. \tag{53}$$

Although we know, of course, that Newton's law is not applicable in quantum mechanics, it is convenient to consider the coefficient of the force in (53) as the inverse of an *effective mass* $\mathbf{m}^*$:

$$\frac{1}{\mathbf{m}^*} = \frac{1}{h^2}\frac{d^2E}{dk^2}. \tag{54}$$

We simply mean by this that (53) allows us to visualize the electron moving in the lattice very much as if it were a classical particle of mass $\mathbf{m}^*$.

It can readily be seen that for free electrons $\mathbf{m}^* = \mathbf{m}$ (since $E = h^2k^2/2\mathbf{m}$). Let us now see what the effective mass is like for the electrons in the linear chain. At the bottom of a nearly free-electron band such as that shown in Fig. 45a, $\mathbf{m}^*$ will be very nearly equal to $\mathbf{m}$, since the $E(k)$

curve hardly differs from the free-electron parabola in this region. However, as we move up along this band the difference between **m** and **m*** becomes more drastic. d^2E/dk^2 measures the curvature of the $E(k)$ curve and it is clear that the curve in the figure has an inflexion point[†] after which d^2E/dk^2 becomes negative. That is, at the top of the band the effective mass of the electrons is negative.

It is a little awkward to visualize the motion of a particle of negative mass, but this can be avoided by introducing a mathematical artifact, called a *hole*, which will be seen to have all the properties of a particle. We define a hole as the absence of an electron: we can say for instance in Fig. 45a that the top of the band is *occupied* by holes. If we remove one electron at a certain point from a Fermi gas, a positive charge will be left. On the other hand, in doing this we have created a hole. A hole, therefore, has positive charge. We shall also see that its mass is equal and opposite in sign to the electron mass. In fact, when an electron moves from A to B we can say that a hole moves from B to A. That is, the holes move in the opposite direction to that of the corresponding electrons: their velocity (and also acceleration) must be equal and opposite to that of the electrons, and the same must be the case for their mass.

If we have electron states with negative mass at the top of a band we can instead talk of hole states with positive mass and this will be convenient since it is easier to visualize the motion of particles of positive mass. When we have a band which is almost full it is particularly useful to interpret electrical conduction as due to the motion of positively charged holes, and one then says that the electrical carriers are holes rather then electrons.

20. The wave functions

If we look at the energy bands in the reduced scheme of Fig. 37, the relevant parts of which are reproduced in Fig. 46, we notice that they behave in two fundamentally different ways. The energy in the first band increases with k, whereas in the second it decreases with k, and so on for the higher bands. (It should be clearly understood that we are still considering nearly free electrons, i.e. bands for which the free-electron

---

† This must necessarily exist if we have to follow the free-electron parabola at the centre of the Brillouin zone and satisfy the condition $dE/dk = 0$ at its edge.

parabola is a reasonable first approximation. The behaviour of the bands in a general case will be more complicated.) It will be useful to obtain some understanding of the origin of this difference by means of a picture that, although crude, will provide an example of some properties of the bands which is important to keep in mind.

Let us consider a chain of hydrogen atoms and take all the z axes in the direction of the chain. As we know, the atomic levels of the free atom are in the order $1s$, $(2s, 2p_z, 2p_x, 2p_y)$, $3s$, etc., where the brackets denote degenerate levels. The first point to notice is that the degeneracy

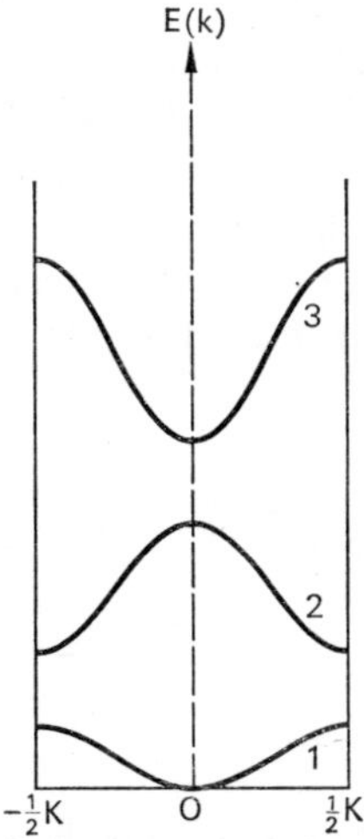

FIG. 46. Different behaviour of bands.

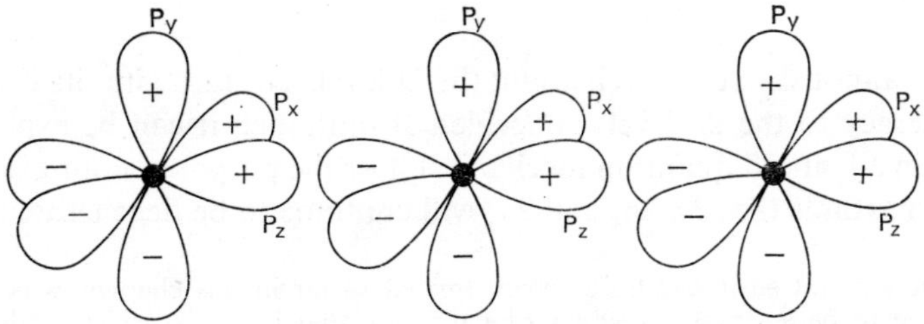

FIG. 47. Orbitals of hydrogen atoms in a linear chain.

$(2p_z, 2p_x, 2p_y)$ is broken when the chain is formed. This follows from Fig. 47, where a few atoms of the chain are shown with the p orbitals in their well known representation: p_x and p_y are still degenerate since a rotation around the chain axis (which is a permitted symmetry operation†) brings one into the other. On the other hand, whereas in the free atom there are permitted rotations that will transform p_z into p_x or p_y so that these functions are degenerate, this is no longer the case in the linear chain, which confirms that p_z is no longer degenerate with the other two p orbitals. We represent this result in Fig. 48. It is quite common in crystals for degeneracies to be broken down in the manner described, a situation which is referred to as the *crystal-field splitting of the degeneracy*. The positioning of the $2p_z$ level below the other two in Fig. 48 must be determined by calculation of the corresponding energy levels, but it can be roughly justified: we see from Fig. 47 that the overlap between two neighbouring $2p_z$ orbitals is larger than for the $2p_x$ or $2p_y$ ones, and it is a well known idea in elementary atomic and molecular structure that the greater such overlaps are, the lower the energy of the corresponding level.

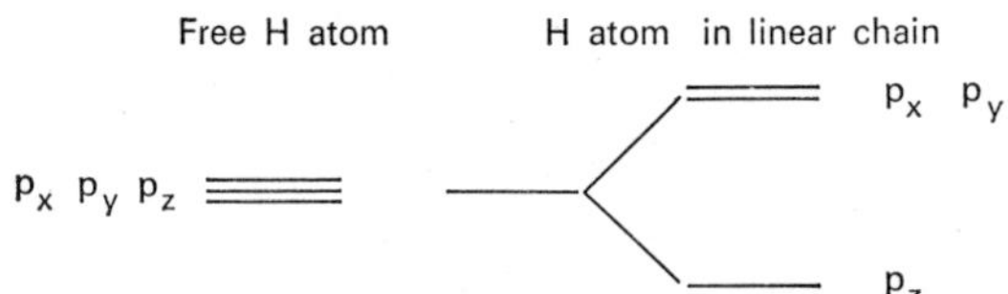

FIG. 48. Splitting of the triply degenerate p level of the hydrogen atom into a doublet and a singlet by the crystal field of the linear chain.

We cannot say very much about the $2s$ level. To start with, its original degeneracy to the $2p$ levels is accidental (although it can be explained away in advanced quantum mechanics). For the purpose of our example we can assume that $2s$, $2p_x$ and $2p_y$ will continue to be degenerate in the

---

† But was not so in the linear chain treated so far in this chapter, which was supposed to be a strictly one-dimensional chain, that is one in which nothing at all existed outside the chain axis (cf. § 14).

linear chain, so that the order of the hydrogen atom levels to be considered is $1s$, $2p_z$, $(2s, 2p_x, 2p_y)$. . . . When we bring N atoms together each level will form a band of N energy levels.

Let us consider the $1s$ band and, since we are prepared to use a rough model, let us assume first that the crystal wave function in one of the band levels looks just like an array of $1s$ functions sitting around each atom (Fig. 49). The situation, however, cannot be quite as simple as this.

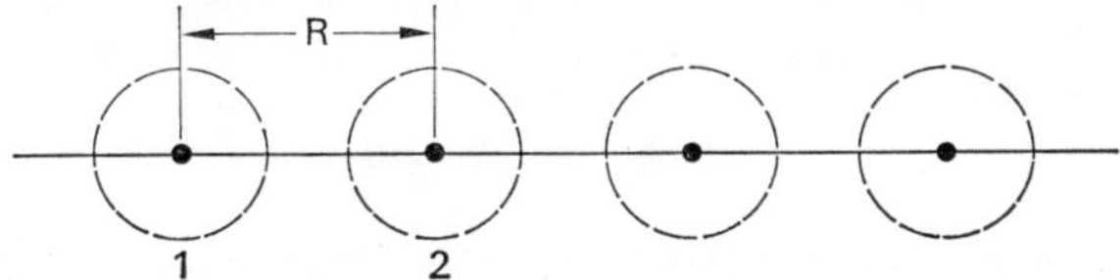

FIG. 49. Hydrogen atoms in $1s$ states in a
linear chain.

The wave function must be a Bloch function which, when we translate the whole crystal from 1 to 2 by the translation **R** must get multiplied by the eigenvalue of R, $\exp(2\pi ikR)$ given by (15). Since R coincides with the lattice constant a, we shall write this eigenvalue in the form $\exp(2\pi ika)$. The general case is a little difficult, but we can easily see what the effect of this factor is for two simple values of k:

$$k = 0 \text{ (centre of Brillouin zone: } \exp(2\pi ika) = 1, \tag{55}$$

$$k = \tfrac{1}{2}K = \tfrac{1}{2}a^{-1}$$
$$\text{(edge of Brillouin zone): } \exp(2\pi ika) = \exp((\pi i)) = -1. \tag{56}$$

Clearly, these relations necessitate some changes in Fig. 49. For them to be satisfied we must now redraw it as shown in Fig. 50.

We know that the energy of a wave function is roughly proportional to the number of its nodes (see § 1.14, iii), which have been counted in Fig. 50. We must therefore expect for the $1s$ band $E(\tfrac{1}{2}K) > E(0)$, i.e. that the energy at the edge of the zone is larger than at the centre. This agrees with the behaviour of band 1 in Fig. 46, which we had to expect would describe the $1s$ band, since $1s$ is the lowest atomic level.

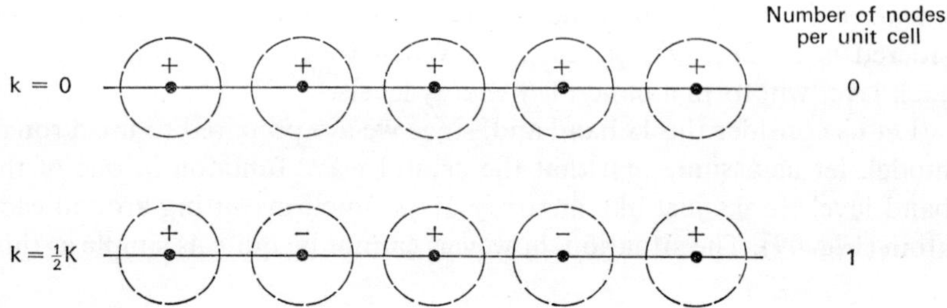

FIG. 50. A rough representation of two Bloch functions of the 1s band. Verify that when a tracing of the wave function for $k = \frac{1}{2}K$ is displaced by the distance R of Fig. 49, the new wave function has changed sign, as required by (56).

In the same manner, we expect band 2 of that figure to correspond to the second, $2p_z$, level. In fact, we must consider again a chain of $2p_z$ orbitals and the numerical factors given in (55) and (56) still apply, which leads to the arrangements shown in Fig. 51. From the number of nodes shown in the figure it follows that $E(0) > E(\frac{1}{2}K)$, so that $k = 0$ is now the maximum and not the minimum of the energy.

We notice that the almost free-electron picture of Fig. 46 gives the second band above the first at the edge of the zone, which gives origin to the energy gap. We can obtain an idea why this arises by looking at Figs. 50 and 51. At the edge of the Brillouin zone both bands have exactly the same number of nodes and can be expected to have similar values

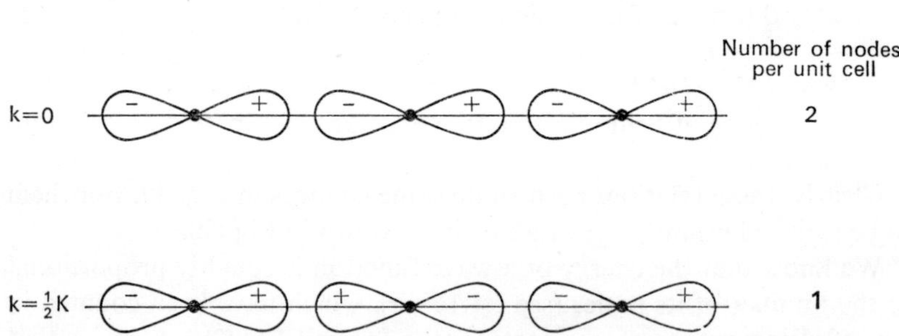

FIG. 51. A rough representation of the Bloch functions of the $2p_z$ band. Verify the property discussed in the legend of Fig. 50. For $k = 0$ there are nodes at the nuclei and midway through them. For $k = \frac{1}{2}K$ there are nodes only at the nuclei.

of the energy. However, for the second, $2p_z$, band the nodes appear at the atoms whereas for the $1s$ band they are at the mid-points between them. Hence, for this value of k, the electrons in the second band are prevented from being near the nuclei where the potential energy is lowest (attraction greatest) so that their energy will be lifted up with respect to the same value of k in the $1s$ band, for which the electrons are allowed to be near the nuclei. [Cf. remark (i) in § 22.2.]

The third band would be triply degenerate, corresponding to the $2s$, $2p_x$, and $2p_y$ levels, all of which can readily be seen to behave, from the point of view of the number of nodes, just as the $1s$ band in Fig. 50. (Such degeneracies, of course, were not permitted in our original, strictly one-dimensional linear chain.)

It is interesting, finally, to observe a property of the $1s$ band. This name must not be taken to mean that for all values of k the wave functions in this band behave exactly like $1s$ orbitals. On the contrary, we see in Fig. 50 that at the edge of the Brillouin zone the wave function has a good deal in common with the $2p_z$ function at the same value of k (Fig. 51). We have already observed that the two pictures have the same number of nodes. In a way, they can be considered to be related by a lateral displacement that shifts the nodes from their position between the nuclei in the $1s$ band to that at the nuclei in the $2p_z$ one. We can say that the $1s$ band has some p-like character (or behaviour) at the edge of the Brillouin zone. This phenomenon, whereby different types of wave functions are mixed in one band, is called *hybridization*, and it is even more important in three-dimensional cases.

21. Lattice with basis

Not all the atoms in the lattice shown in Fig. 52 are translationally equivalent: we can never go from one of the black to one of the white atoms by a translation of the lattice. In practice the atoms of the two different types may or may not be of the same chemical element: they are, nevertheless, always distinct on account of the type of site that they occupy. Such a lattice is called a *lattice with basis* since each unit cell contains more than one atom and the contents of the unit cell is often referred to as the *basis*. This is the pattern the periodic repetition of which generates the whole system.

It should be clearly understood that in so far as the quantization of k values is concerned, there is no difference whatsoever between this lattice and the one shown in Fig. 22 (p. 62). The k space is entirely blind to the contents of the unit cell, and the whole of our previous theory for the Bloch functions is valid for lattices with basis without any formal changes at all. The only conceptual difference appears in the nature of the cell function which will no longer tend in the limit $a \to \infty$ to a single atomic function (since at this limit the basis still holds together), but rather to some molecular orbital.

FIG. 52. A lattice with basis.

Just as we had before, if there are N unit cells in the lattice with basis (Fig. 52) the Brillouin zone contains N states (or N values of k) or, with spin, $2N$ states. This is the same as two states (including spin) *per unit cell* which is a completely general result. From it, it follows that in the case of Fig. 52 the Brillouin zone will contain one state per atom.

The reader must be careful about this: very often one says that the Brillouin zone contains two states per atom but this is a loose description, which is strictly valid for simple lattices only, that is lattices without bases. It is hardly more troublesome and a great deal safer to say always that the *Brillouin zone contains two states per unit cell of the crystal lattice, spin included.* Mistakes and ambiguities also arise here very often through lack of an explicit statement as to whether spin has been counted or not.

In the rest of this chapter we shall develop some computational techniques that will allow us, in principle, to compute in detail the shape of bands such as those given in Fig. 46. That is, for a given weak periodic potential we shall be able to compute the departure of the energy bands from the free-electron values, work which we have so far done in a qualitative way only. In doing so, we shall obtain alternative analytical proofs of some of our most important results: the existence of energy gaps, the $E(k) = E(-k)$ degeneracy and the prohibition of band crossings. These alternative proofs will provide a new and useful insight into these properties. Also, some useful tools will be given that will be further developed later on for three-dimensional crystals.

● 22. Exercise: the energy gap

Consider the normalized free-electron wave functions

$$\psi_k(x) = L^{-\frac{1}{2}} \exp(2\pi i k x),$$

$$\psi_{-k}(x) = L^{-\frac{1}{2}} \exp(-2\pi i k x), \quad \text{with} \quad k = \kappa/L, \quad \text{for integral} \quad \kappa. \tag{57}$$

(Cf. **2.24**.) We know that $\psi_k(x)$ and $\psi_{-k}(x)$ are degenerate eigenfunctions of the energy for free electrons.

Show that the degeneracy between $\psi_k(x)$ and $\psi_{-k}(x)$ is not removed by a weak periodic potential except when k reaches the edge of the Brillouin zone ($k = \frac{1}{2}K_m$).

1. METHOD

1. *Trial wave function*

Since the potential is weak, the free-electron eigenfunctions $\psi_k(x)$ and $\psi_{-k}(x)$ are reasonable approximations to the correct wave function. Write a trial wave function

$$\Psi = c_1\psi_k(x) + c_2\psi_{-k}(x) \tag{58}$$

as in the exercise of § **1**.18 and identify $\psi_k(x)$ and $\psi_{-k}(x)$ with the functions f_1 and f_2 respectively in that exercise.

A clearer picture of the meaning of this approximation will be obtained in § 23.

2. *Elements of the secular determinant*

In order to write down the secular determinant (**1**.77) remember that $S_{11} = S_{22} = 1$, $S_{12} = 0$ since, from the result of the exercise in § **2**.5, the functions (57) are orthonormal.

You should also prove that $H_{22} = H_{11}$. In order to do this write

$$H_{22} = \int \psi^*_{-k}\mathbf{H}\psi_{-k}\,d\tau \tag{59}$$

$$= \int \psi_k\mathbf{H}\psi^*_k\,d\tau \tag{60}$$

$$= \int \psi_k\mathbf{H}^*\psi^*_k\,d\tau = \int (\mathbf{H}\psi_k)^*\psi_k\,d\tau \tag{61}$$

$$= \int \psi^*_k\mathbf{H}\psi_k\,d\tau \tag{62}$$

$$= H_{11}. \tag{63}$$

E

Here, in (60), we use the fact that, from (57), $\psi_{-k} = \psi_k^*$. In (61) we use first the reality property of $\mathbf{H}$ ($\mathbf{H} = \mathbf{H}^*$) and then a trivial rearrangement. In (62) the hermitian property of $\mathbf{H}$ is used.

3. *The secular determinant*

It now follows from (1.77) and (63) that the secular determinant takes the form

$$\begin{vmatrix} H_{11}-\varepsilon & H_{12} \\ \\ H_{21} & H_{11}-\varepsilon \end{vmatrix} = 0. \tag{64}$$

From Exercise 4, § 1.19, $H_{21} = H_{12}^*$ and equation (64) can be written in the form $(H_{11}-\varepsilon)^2 = H_{12}H_{12}^*$, whence $H_{11}-\varepsilon = \pm|H_{12}|$, where the vertical bars denote moduli of complex numbers. Therefore, the two roots of (64), which are the two possible values of the energy for the weak field, are

$$\varepsilon_\pm = H_{11} \pm |H_{12}|. \tag{65}$$

Notice that whereas before $\psi_k(x)$ and $\psi_{-k}(x)$ corresponded to a single energy eigenvalue, the new wave function defined by (58) now gives place to two eigenvalues, as long as $H_{12} \neq 0$.

4. *The wave functions*

To obtain the wave functions that correspond to ε_+ and ε_- you must find the coefficients c_1 and c_2 in (58). In order to do this remember (see § 1.18) that the secular determinant (64) is the determinant of the coefficients of the homogeneous system of linear equations of which c_1 and c_2 are the unknowns. Thus the first of these equations is

$$c_1(H_{11}-\varepsilon)+c_2 H_{12} = 0. \tag{66}$$

For simplicity, we shall consider the case when H_{12} is real† and we

---

† This is always the case when there is a centre of symmetry (see Problem at the end of this section). When this is not the case H_{12} should be written as $|H_{12}|\exp{(i\phi)}$, where ϕ is the phase of the complex number. It follows from (66) and (65) that $c_1 = \pm c_2 \exp{(i\phi)}$. In order to write Ψ_+ and Ψ_- as in (70) and (71), take out $\exp{(\frac{1}{2}i\phi)}$ as common factor, the effect of which is: (i) to provide an irrelevant phase factor in the wave function; (ii) to replace x in (70) and (71) by $x+\alpha$, where $\alpha = \phi/(4\pi k)$. This change of phase in x merely displaces the nodes of the standing waves (70) and (71). (See the remark below, in § 2 (i) of this section.)

shall assume that it is positive. Then, on introducing (65) into (66),

$$c_1(\mp H_{12}) + c_2 H_{12} = 0. \tag{67}$$

Hence, there are two solutions for c_1 and c_2:

$$c_1 = c_2 \quad \text{for} \quad \varepsilon_+ = H_{11} + H_{12}, \tag{68}$$

$$c_1 = -c_2 \quad \text{for} \quad \varepsilon_- = H_{11} - H_{12}. \tag{69}$$

When the values of the coefficients that appear in (68) and (69) are introduced in (58) you will obtain the two wave functions that correspond to ε_+ and ε_-, which can be called Ψ_+ and Ψ_- respectively. Since c_2 will appear as a constant factor in these expressions it can be taken to be unity. When this is done and the values (57) are used for $\psi_k(x)$ and $\psi_{-k}(x)$ in (58), you will find

$$\Psi_+(x) = \cos 2\pi k x, \tag{70}$$

$$\Psi_-(x) = \sin 2\pi k x, \tag{71}$$

which are standing waves, whereas the original functions (57) were travelling waves.

5. *The case* $H_{12} = 0$

When this relation is satisfied you must notice two things. First, from (65), $\varepsilon_+ = \varepsilon_-$, so that the two energy eigenvalues are equal and the degeneracy persists. Secondly, from (67), it turns out that the solutions $c_1 = c_2$ and $c_1 = -c_2$ are no longer possible: the original wave functions are not altered. That is, when $H_{12} = 0$ the degeneracy is not removed and the wave functions are the original travelling waves.

6. $H_{12} = 0$ *for all* $k \neq \tfrac{1}{2}K_m$

Clearly, this result will have to be proved in order to complete the exercise. This is the most difficult step. We shall give here a proof of it by means of symmetry considerations and an alternative analytical proof in § 24.

Consider the general term

$$H_{kk'} = \int \psi_k^* \mathbf{H} \psi_{k'} \, dx. \tag{72}$$

Write

$$\mathbf{H} = \mathbf{H}_0 + V(x), \tag{73}$$

where $\mathbf{H}_0$ is the free-electron hamiltonian and V is the weak periodic potential. From (72),

$$H_{kk'} = \int \psi_k^* \mathbf{H}_0 \psi_{k'} \, dx + \int \psi_k^* V(x) \psi_{k'} \, dx. \tag{74}$$

Since $\psi_{k'}$ is a free-electron eigenfunction,

$$\mathbf{H}_0 \psi_{k'} = E_{k'} \psi_{k'}, \tag{75}$$

where $E_{k'}$ is the free-electron eigenvalue. Therefore, with the definitions

$$S_{kk'} = \int \psi_k^* \psi_{k'} \, dx, \tag{76}$$

$$V_{kk'} = \int \psi_k^* V(x) \psi_{k'} \, dx, \tag{77}$$

(74) gives

$$H_{kk'} = E_{k'} S_{kk'} + V_{kk'}. \tag{78}$$

Consider the term $V_{kk'}$. Clearly this term, like $H_{kk'}$, is an energy quantity that cannot be altered if a symmetry operation is applied on the periodic lattice. That is, we must have

$$\mathbf{R} V_{kk'} = V_{kk'}, \tag{79}$$

when $\mathbf{R}$ is one of the translation operators. The left-hand side of (79) is the integral (77) after everything in the integrand has been transformed by the translation operator $\mathbf{R}$:

$$\mathbf{R} V_{kk'} = \int (\mathbf{R}\psi_k)^* \mathbf{R} V(x) \mathbf{R}\psi_{k'} \, dx. \tag{80}$$

Since $V(x)$ is periodic, $V(x+R) = V(x)$, whence

$$\mathbf{R} V(x) = V(x+R) = V(x). \tag{81}$$

From (57),

$$\mathbf{R}\psi_k(x) = \psi_k(x+R) = \exp(2\pi i k R)\psi_k(x). \tag{82}$$

On introducing (81) and (82) in (80),

$$\mathbf{R}V_{kk'} = \exp[2\pi i(k'-k)R]V_{kk'}, \tag{83}$$

from which, in order to satisfy (79), it follows that the condition

$$(k'-k)R = m, \quad \text{for } m \text{ integral,} \tag{84}$$

has to be satisfied. We can take R to be the translation a that corresponds to the operator $\mathbf{R}_1$, so that

$$k'-k = ma^{-1} = mK = K_m. \tag{85}$$

(See § 7.2.) If (85) is not satisfied the left- and right-hand sides of (79) cannot be equal unless $V_{kk'} = 0$. This means that $V_{kk'} \neq 0$ if and only if k and k' differ by K_m. For the case in which we are interested, that is $V_{k,-k}$, this condition requires that $-k-k = K_m$ or $k = -\frac{1}{2}K_m$. Also, in the case under consideration $k \neq k'$, so that $S_{kk'} = 0$ (cf. exercise in § 2.5). Therefore, from (78), $H_{12} = H_{k,-k} = V_{k,-k}$, and H_{12} will be zero unless $k = -\frac{1}{2}K_m$. (Notice that this condition is equivalent to $k = \frac{1}{2}K_m$ by the periodicity in k.) This completes the proof.

7. *Results*

From (78), since $S_{kk} = 1$ and $S_{k,-k} = 0$:

$$\left.\begin{aligned} H_{11} &= H_{kk} = E_k + V_{kk}, \\[2ex] H_{12} &= V_{k,-k}, \end{aligned}\right\} \tag{86}$$

so that, from (65),

$$\varepsilon_{\pm} = E_k + V_{kk} \pm V_{k,-k}. \quad (k = \tfrac{1}{2}K_m). \tag{87}$$

In this equation, E_k is the free-electron energy that corresponds to $k = \frac{1}{2}K_m$ and $V_{kk} = L^{-1}\int_0^L V(x)dx$ [see (77) and (57)] is the average of the potential energy over the lattice, which can conveniently be taken as the energy origin, so that $V_{kk} = 0$. $\pm V_{k, -k}$ is therefore the correction to the free-electron eigenvalue at the edge of the Brillouin zone. From (57),

$$V_{k, -k} = V_{\frac{1}{2}K_m, -\frac{1}{2}K_m} = L^{-1} \int_0^L \exp(-2\pi i K_m x) V(x) dx. \qquad (88)$$

2. Remarks

(i) You can see why Ψ_+ and Ψ_- from (70) and (71) are no longer degenerate when $k = \frac{1}{2}K = \frac{1}{2}a^{-1}$, in the periodic field. In fact, for this value of k

$$\Psi_+ = \cos(\pi x a^{-1}), \quad \Psi_- = \sin(\pi x a^{-1}), \qquad (89)$$

whence

$$\Psi_+ = 0, \quad \Psi_- = 1, \quad \text{for} \quad x = \tfrac{1}{2}a,$$

$$\Psi_+ = 1, \quad \Psi_- = 0, \quad \text{for} \quad x = 0,$$

which means that the node of Ψ_+ is at the centre of the cell, whereas that for Ψ_- is at the origin ($x = 0$). If the latter is the region where the potential is lowest (for instance, because there is an atom at that point, so that the nuclear attraction is a maximum) Ψ_- is energetically unfavourable, since the chances of finding the electron at the region of lowest potential are lowest.

(ii) We have proved that the degeneracy of the free-electron eigenfunctions ψ_k and ψ_{-k} breaks down when $k = \frac{1}{2}K$. However, we have repeatedly said that $E(k) = E(-k)$ (see § 9), so that it might appear that we are in contradiction with our previous work and the reader may be wondering what all this is about.

We must remember, however, that the sequence of values of k in a Brillouin zone is not strictly symmetric: for any given k in the Brillouin zone the value $-k$ belongs to the same Brillouin zone, *except* when $k = \frac{1}{2}K$. It follows, in fact, from (41) that the sequence of values of k in the first Brillouin zone is

$$k = -\tfrac{1}{2}K, \quad \left(-\tfrac{1}{2}K + \frac{K}{N}\right), \quad \left(-\tfrac{1}{2}K + 2\frac{K}{N}\right), \ldots, \left(\tfrac{1}{2}K - \frac{K}{N}\right). \qquad (90)$$

In (90), all values of k appear in pairs $(-k, k)$, except the very first one.

What we have proved in this exercise is this: for all values of k in the Brillouin zone listed in (90), except the first, the free-electron degeneracy $E(k) = E(-k)$ is preserved in the periodic field. The pair of free-electron eigenfunctions $\psi_{-\frac{1}{2}K}$, $\psi_{\frac{1}{2}K}$ correspond in the weak field, from (90), to the single value of $k = -\frac{1}{2}K$. Now, for this value of k, since the original degeneracy of $\psi_{-\frac{1}{2}K}$ and $\psi_{\frac{1}{2}K}$ is broken, there correspond two values of the energy ε_+ and ε_- and two wave functions. These two values of the energy which, from (87) are separated by $2V_{k,\,-k}$, correspond to the values of the energy for the first and second bands at $k = -\frac{1}{2}K$. $2V_{k,\,-k}$ is the energy gap at the edge of the Brillouin zone which we have found before in a different way.

(iii) Finally, the reader should observe that this exercise provides an alternative proof of the degeneracy of the Bloch functions that belong to k and $-k$ (cf. § 9).

PROBLEM. Show that if the potential is centro-symmetrical, that is, if $V(x) = V(-x)$, H_{12} in (64) is real.

Hints. $H_{12} = H_{kk'}$ is given by (78). The first term on the right-hand side is real. Introduce the wave functions (57) in $V_{kk'}$ from (77). Then,

$$V_{kk'} = L^{-1} \int \exp[2\pi i(k'-k)x] V(x) dx.$$

This integration should be carried out for x from 0 to L, but it is convenient to take the origin at the centre of the chain, so that $-\frac{1}{2}L \leqslant x \leqslant \frac{1}{2}L$. Decompose the exponential in its cosine (even) and sine (odd) parts. For each integral apply the result of Exercise 6, § **1**.19, whence the sine (imaginary) part of the integral vanishes.

Remark. Whereas a simple linear chain is always centro-symmetrical, this is not so for one with a basis (§ 21). In this case, H_{12} will in general be complex.

● 23. The nearly free-electron approximation

We computed in § 22 the energies of two successive bands at a single value of k, namely the edge of the first Brillouin zone, and we thereby obtained the energy gap. This work suggests a method to obtain the energy eigenvalues for all bands at any value of k, which we are now going to discuss. When we do this, the work of § 22 will be seen in the light of a more general framework, and it will be a great deal easier to appreciate.

If we wish to compute the energy eigenvalues for a given k in a weak periodic potential we can use the general linear variational expression of § **1**.18 in order to write the following trial wave function:

$$\Psi_k(x) = \sum_{m=0}^{M} c_m \psi_{k+K_m}(x), \tag{91}$$

where M is some integer, $\psi_k(x)$ is the free-electron eigenfunction defined in (57), and $\psi_{k+K_m}(x)$ is obtained by substituting $k+K_m$ for k in it.

The form of expansion (91) is far from obvious at this stage, and we shall now give the reasons that account for the various details of it.

(i) We know that the free-electron eigenfunctions are good first order approximations in a weak periodic potential so that it is natural to use them in a trial variational wave function.

(ii) All the functions that appear on the right-hand side of (91) belong to the same value of k in the periodic field, since $k+K_m$ and k are identical in this case (although the free-electron eigenfunctions ψ_k and ψ_{k+K_m} are not so). This justifies the assignment of the wave function on the left of (91) to the value k.

(iii) We saw in the exercise of § 22 that if the functions used in the variational expression do not have off-diagonal matrix elements in the determinant, no change whatever is experienced in the energy levels, so that it is a waste of time to include such functions in the expansion. It can readily be seen that this is also the case when larger expansions, such as (91), are used. Accepting now this result and calling for simplicity k and k' two values of k that appear in (91), let us obtain the corresponding matrix element $H_{kk'} - \varepsilon S_{kk'}$ [see (72) and (76)]. Since the free-electron eigenfunctions are orthogonal (i.e. $S_{kk'} = 0$), this matrix element reduces into $H_{kk'}$. In its turn, from (78) and on account of the same property, $H_{kk'} = V_{kk'}$. Therefore, the only wave functions that are worth including in (91) are those for which $V_{kk'} \neq 0$ and from (85) this is the case only when $k' = k+K_m$, which is exactly the form of the values of k included in the summation in (91).

Suppose now that, for the sake of an example, we take $M = 4$ in (91). We shall have a 5×5 secular determinant, that is a fifth-degree equation in ε from which five eigenvalues will be obtained. For each eigenvalue, in the manner of § 22, we obtain a set of coefficients $c_0, c_1, \ldots, c_4$, i.e. a wave function $\Psi_k(x)$. Therefore, we obtain five eigenvalues and five eigenfunctions for the first five bands of the metal at the value of k chosen. Since k can be varied throughout the Brillouin zone, we now have a useful method of approximation that will allow us to compute in detail band systems like the one depicted in Fig. 46.

We can now go back to the work of § 22. We took the functions $\psi_k(x)$ and $\psi_{-k}(x)$ in (58) because we were interested to see whether their degeneracy would be split by the weak field. However, from (91) we know that if we take only two functions in the expansion ($M = 1$) the only ones worth taking are ψ_k and ψ_{k+K}. For a general value of k we shall then obtain the corresponding eigenvalues of the first two bands. For $k = -\frac{1}{2}K$ we obtain those at the edge of the Brillouin zone, as we did in the exercise of § 22. On the other hand, it follows at once from expansion (91), again in the case of two functions only, that the pair $k,\ -k$ will work only when $-k = k + K$, i.e. when $k = -\frac{1}{2}K$, which is the edge of the Brillouin zone.

It should be appreciated that the numerical results of the nearly free-electron approximation depend on the determination of the coefficients $V_{kk'}$ defined in (77), when $k' = k + K_m$. [See, for instance, (87).] These terms are called the matrix elements of the potential and their determination will be discussed in the next section.

● 24. Fourier coefficients of the potential

We have seen that the nearly free-electron approximation depends crucially on the fact that the matrix elements of the potential $V_{kk'}$ defined in (77) vanish if $k' \neq k + K_m$. We proved this in § 22 by using symmetry arguments and we shall now provide an alternative analytical proof, which will allow us to clarify the meaning of these matrix elements.

Since we have to use the *Fourier series* we shall first revise it briefly.

E*

1. Fourier Expansion of the Potential

Suppose that the functions $\varphi_m(x)$ for m an integer from $-\infty$ to ∞ are orthonormal (see the exercise in § 2.5). That is,

$$\int_a^b \varphi_{m'}^*(x)\varphi_m(x)dx = \delta_{mm'}, \tag{92}$$

where a and b are some values of the variable x.

Given an arbitrary function $f(x)$ we try to write it in terms of the functions $\varphi_m(x)$:

$$f(x) = \sum_{m=-\infty}^{\infty} c_m \varphi_m(x). \tag{93}$$

This is an infinite series, and the possibility of the expansion (93) will depend on the convergence of the series, with which we shall not be concerned here. Instead, we shall simply observe that, if (93) is meaningful, the coefficients c_m are then very simple to obtain: Multiply both sides of (93) by $\varphi_{m'}^*(x)$ and integrate. On using (92),

$$\int_a^b \varphi_{m'}^*(x)f(x)dx = \sum_m c_m \delta_{mm'} = c_{m'}, \tag{94}$$

which gives the value of each coefficient in (93) in terms of a simple integration which is always possible, at least numerically.

One condition must be noticed: clearly, for expansion (93) to be at all possible, the function $f(x)$ must satisfy the same boundary conditions as the functions $\varphi_m(x)$ at the ends a and b of the domain over which the latter are defined.

We shall use these ideas in order to expand the metal potential. This is periodic in x with period a:

$$V(0) = V(a). \tag{95}$$

On the other hand, the functions

$$\varphi_m(x) = a^{-\frac{1}{2}} \exp(2\pi imxa^{-1}) \tag{96}$$

satisfy exactly the same periodicity and it can also be readily seen that they are orthonormal over the domain $0 \leqslant x \leqslant a$. (This can be proved at once, when it is realized that they coincide with the functions $\psi_i(x)$ of the exercise in § 2.5, if L is replaced by a and the integer κ_i by m.)

Since $ma^{-1} = K_m$, we can rewrite the functions (96) in the more compact form

$$\varphi_m(x) = a^{-\frac{1}{2}} \exp(2\pi i K_m x). \tag{97}$$

In order to expand the potential $V(x)$ we substitute it for the function $f(x)$ in (93), on using the functions (96). For convenience, we call now the coefficients V_m rather than c_m, and we obtain

$$V(x) = a^{-\frac{1}{2}} \sum_m V_m \exp(2\pi i K_m x), \tag{98}$$

where, from (94),

$$V_m = a^{-\frac{1}{2}} \int_0^a \exp(-2\pi i K_m x) V(x) dx. \tag{99}$$

2. The matrix elements of the potential

We shall now consider the matrix element $V_{kk'}$ from (77),

$$V_{kk'} = \int \psi_k^*(x) V(x) \psi_{k'}(x) dx, \tag{100}$$

with the free-electron eigenfunctions (57),

$$\psi_k(x) = L^{-\frac{1}{2}} \exp(2\pi i k x), \qquad k = \kappa/L. \tag{101}$$

The integration in (100) must be performed over the whole domain of definition of the functions (101), i.e. from 0 to L. Therefore

$$V_{kk'} = L^{-1} \int_0^L \exp[2\pi i(-k+k')x] V(x) dx, \tag{102}$$

and, on introducing (98), we obtain

$$V_{kk'} = L^{-1} a^{-\frac{1}{2}} \sum_m V_m \int_0^L \exp[2\pi i(-k+k'+K_m)x] dx. \tag{103}$$

The integral in (103) is a constant times $\exp\left[2\pi i(-k+k'+K_m)x\right]$, which, on introducing $k = \kappa L^{-1}$ and $K_m = ma^{-1} = mNL^{-1}$ (for $L = Na$), can be rewritten as

$$\exp\left[2\pi i(-\kappa+\kappa'+Nm)L^{-1}x\right].$$

This exponential equals unity for $x = 0$ and, since $(-\kappa+\kappa'+Nm)$ is an integer, for $x = L$, so that the integral in (103) vanishes. This deduction is valid except when the integrand in (103) equals unity, i.e. except when

$$-k+k'+K_m = 0, \tag{104}$$

which is the condition for the integral, and therefore $V_{kk'}$, not to vanish, as we already found in (85) of § 22. (There is, in fact a change in the sign of m from one equation to the other, but this does not matter since in either case m is an arbitrary integer, positive or negative.)

For a given pair k, k' there will be only one value of m that satisfies (104). The only integral that does not vanish on the right-hand side of (103) is equal to L. Therefore

$$V_{kk'} = a^{-\frac{1}{2}}V_m \tag{105}$$

or

$$V_{k,\ k-K_m} = a^{-\frac{1}{2}}V_m, \tag{106}$$

where V_m is obtained from (99). These integrations can easily be performed if the potential is known. Nevertheless, since the detailed form of the potential is difficult to determine in a real metal, its matrix elements (or, what is the same from 105, the Fourier coefficients of the potential) are often used as adjustable parameters which are determined from values of energy gaps. (See Problem 3 below.)

<h3 align="center">● Problems</h3>

1. Derive (106) directly from (88).

Hints. In (88), $L = Na$, where N is the number of unit cells in the chain. Also, the integral from 0 to L is, from periodicity, N times the integral from 0 to a. Substitute the value of the latter integral from (99).

2. Show that, if there is a centre of symmetry,

$$V_m = V_{-m} = a^{-\frac{1}{2}} \int_0^a \cos(2\pi mxa^{-1})V(x)dx.$$

Hint. Remember that if we take the origin of the unit cells at the nuclei $V(x) = V(-x)$ which will result in the cancellation of the imaginary part of (98).

3. Show, from the results of § 22 and (105) or (106), that the energy gap at the edge of the *m*th Brillouin zone is $2a^{-\frac{1}{2}}V_m$. (That is, the energy gap is the *m*th Fourier coefficient of the potential, except for a constant.)

4. Show that the energy gap at the edge of the first Brillouin zone is $2a^{-1} \int_0^a \cos(2\pi xa^{-1})V(x)dx$. Hence, verify that there is no energy gap for a constant potential.

5. Show that two bands of the linear lattice cannot cross.

Hints. Consider two bands 1 and 2 that cross at k. That is, the Bloch functions $\varphi_k^1(x)$ and $\varphi_k^2(x)$ are such that $E^1(k) = E^2(k)$ for the particular value of k under consideration. Write a variational function $c_1\varphi_k^1(x) + c_2\varphi_k^2(x)$. Proceeding along the lines of § 22 you will see that the crucial step is to show that the matrix element $\int [\varphi_k^1(x)]^* V(x)\varphi_k^2(x)dx \neq 0$. You can do this exactly as in § 22.1.6. This treatment, of course, confirms the result of § 14.

CHAPTER 4

Bloch Functions and Brillouin Zones in Three Dimensions

1. Crystal periodicity

Very few changes are necessary to progress from the treatment of the linear chain in Chapter **3** to that of a three-dimensional crystal. As most often with three-dimensional problems, it is very convenient to use vector notation, which we shall develop in the first few sections of this chapter. To make a smooth connection with Chapter 3 we shall start by rewriting in vector form some of the expressions that we used in it.

The periodicity of the linear chain depends on the single quantity a, the lattice constant, along the x axis. In Chapter 3 we did not discuss the units in which a and x are measured and we must now do this. A little care is necessary to obtain the right result. The main point is that, on the one hand, the lattice constant a must be measured in the standard units of crystallography, i.e. in Ångstroms (Å). Since a is no more than a particular value of the variable x, this variable must also be measured in Ångstrom units. On the other hand, it is a great deal simpler to represent the position r of a point along the chain not in terms of x (measured in Å) but in units of a, i.e. by means of the dimensionless variable x/a. In preparation for the passage to three dimensions we define a vector $\mathbf{a}$ along the x axis. $\mathbf{a}$ is *not* of unit modulus. Rather, $|\mathbf{a}| = a$ measured in Å. The position of a point along the x axis, now represented with the vector $\mathbf{r}$, will be given by $\mathbf{r} = (x/a)\mathbf{a}$. (Notice that when $x = a$ this formula gives the correct value $\mathbf{r} = \mathbf{a}$.)

The vector $\mathbf{a}$ is called a *primitive translation vector* since a translation by it will leave the crystal invariant. (It corresponds, in fact, to the displacement R_1 by a which we used in Chapter **3**, where we saw that all displacements effected by translations were multiples of R_1. In the present

128

notation all displacements $R_m = ma$ can be written straight away as $m\mathbf{a}$ since the integer m is, of course, dimensionless.)

In three dimensions, this periodicity exists along three axes x, y, z on which the basic periods are the lengths a, b, c respectively, normally measured in Ångstrom units. We define along these directions corresponding *primitive vectors* $\mathbf{a}, \mathbf{b}, \mathbf{c}$ (see Fig. 53) such that $|\mathbf{a}| = a$, $|\mathbf{b}| = b$,

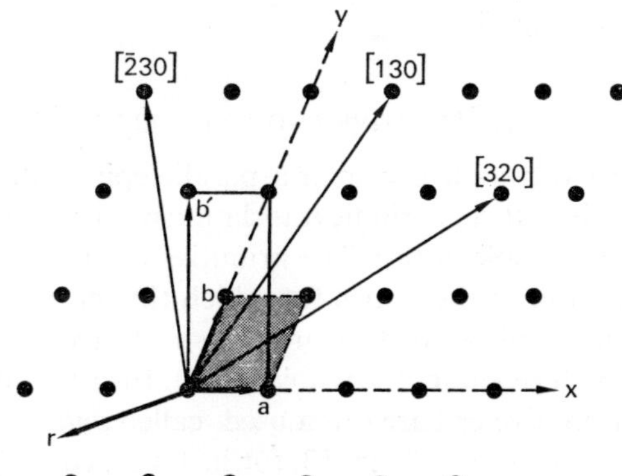

Fig. 53. A translational lattice. The z axis is perpendicular to the plane of the drawing. Three vectors *of* the translational lattice are shown, in the notation of (2). [2̄30] is the vector with component -2 along x, 3 along y, and 0 along z. $\mathbf{r}$ is a vector *in* the lattice, as defined in (1). The primitive cell is shaded and a rectangular unit cell is shown.

$|\mathbf{c}| = c$. (It should be noticed that, in general, $\mathbf{a}, \mathbf{b}, \mathbf{c}$ will neither be unit nor orthogonal.) The position of a point in the crystal will be given by the vector

$$\mathbf{r} = \frac{x}{a}\mathbf{a} + \frac{y}{b}\mathbf{b} + \frac{z}{c}\mathbf{c} = r_x\mathbf{a} + r_y\mathbf{b} + r_z\mathbf{c}, \tag{1}$$

where the dimensionless components r_x, r_y, r_z are given in units of a, b, and c respectively.

A lattice translation, which in one dimension is $\mathbf{R}_m = m\mathbf{a}$, will take the form

$$\mathbf{R}_{mnp} = m\mathbf{a} + n\mathbf{b} = p\mathbf{c} = [m, n, p], \tag{2}$$

where m, n, p, are any integers. Unless more detail is required we shall often write $\mathbf{R}_{mnp}$ as $\mathbf{R}$. The set of all the vectors $\mathbf{R}_{mnp}$ for all integral values of m, n, p, forms a *lattice* which is called the *translational lattice*. The vectors $\mathbf{R}_{mnp}$ themselves (m, n, p integral) are called *vectors of the translational lattice.*† They are such that their heads and tails coincide with lattice points (see Fig. 53).

1. PRIMITIVE AND UNIT CELLS

The primitive vectors $\mathbf{a}, \mathbf{b}, \mathbf{c}$ span a parallelepiped that is called the *primitive cell* of the lattice. Periodic translation of this cell by the primitive vectors covers the whole lattice. The primitive cell is chosen so that it is the smallest portion of the lattice for which this property is valid and, accordingly, the primitive vectors are the shortest vectors the periodic repetition of which generates the whole lattice. In crystallography, larger cells than the primitive cell are often used, called *unit cells* and spanned by *unit vectors*. The lattice of Fig. 53, for instance, could be described in terms of the unit vectors $\mathbf{a}$ and $\mathbf{b}'$ as a rectangular lattice with basis (since each unit cell contains two lattice points, namely the one at the centre and a quarter of each of the four points at the vertices).

In dealing with one-dimensional lattices in Chapter 3 we made no distinction between unit and primitive cells, since this is hardly necessary in such cases: strictly speaking all "unit" cells used in Chapter 3 (whether in direct or reciprocal space) are primitive.

2. Bloch functions in three dimensions

From (3.16) and (3.19) we can write the Bloch function in one dimension, along the x axis, in the following form:

$$\varphi_k^j(x) = \exp\left(2\pi i \, \frac{\kappa x}{Na}\right) u_k^j(x), \tag{3}$$

† The reader should pay attention to the particle "of", the meaning of which is quite specific to the case when m, n, p are integral. When this is not the case, as for the vector $\mathbf{r}$ in Fig. 53, the vector is a vector *in* the translational (or crystal) lattice.

where κ is an integer in the range $1 \leqslant \kappa \leqslant N$ and N is the number of unit cells along x.

Since $x/a = r_x$, we can rewrite (3) and introduce a notation that will allow us to handle the two other dimensions as well:

$$\varphi^j_{k_x}(x) = \exp{(2\pi i k_x r_x)}u^j_{k_x}(x), \tag{4}$$

$$\varphi^j_{k_y}(y) = \exp{(2\pi i k_y r_y)}u^j_{k_y}(y), \tag{5}$$

$$\varphi^j_{k_z}(z) = \exp{(2\pi i k_z r_z)}u^j_{k_z}(z). \tag{6}$$

Along each of the x, y, z directions we use the periodic boundary conditions of Born and von Kármán, so that k_x, k_y, k_z are quantized as in (3.19):

$$k_x = \kappa_x/N_x, \quad k_y = \kappa_y/N_y, \quad k_z = \kappa_z/N_z, \tag{7}$$

although a change of notation should be noticed since, before, $k = \kappa/Na$ along the x axis, whereas we have now shifted the denominator a onto x so as to introduce r_x. In (7), $N_i(i = x, y, z)$ are the number of unit cells in the corresponding directions and κ_i are integers in the range $1 \leqq \kappa_i \leqq N_i$ [see (3.26)] or any equivalent range defined by periodicity (see § 3.11).

The functions (4), (5), (6) are eigenfunctions of the translations by **a**, **b**, and **c** respectively: we must now form eigenfunctions of the general translations of the lattice, which are of the form (2). We can repeat exactly the argument used in § 2.6 and, just as there, the required function is the product of the functions (4), (5), and (6), which we write in an obvious notation:

$$\varphi^j_{k_x k_y k_z}(x, y, z) = \exp{[2\pi i(k_x r_x + k_y r_y + k_z r_z)]}u^j_{k_x k_y k_z}(xyz). \tag{8}$$

[The reader should observe that the exponent in (8) is the same as that in (2.27), except for a small change of notation.]

Considerable simplification can be gained by introducing vector notation in (8), particularly since the term $k_x r_x + k_y r_y + k_z r_z$ in the exponent appears to be of the form of a scalar product between two vectors. We shall now discuss this.

3. Scalar product and reciprocal vectors

In this and the following section we shall develop some vector methods that are essential to the understanding of band theory, in particular as regards the concept of the Brillouin zone in three dimensions.

The *scalar product* of two vectors $\mathbf{u}$ and $\mathbf{v}$ is the *number* $\mathbf{u} \cdot \mathbf{v} = |\mathbf{u}|\,|\mathbf{v}| \cos \theta$, where $|\mathbf{u}|$ and $|\mathbf{v}|$ are the lengths (or moduli) of the corresponding vectors and θ is the angle they span. It follows that (i) if two vectors are perpendicular (*orthogonal*) $\mathbf{u} \cdot \mathbf{v} = 0$, (ii) $\mathbf{u} \cdot \mathbf{u} = |\mathbf{u}|^2$.

It is customary in vector work to express a vector $\mathbf{u}$ in terms of its components u_x, u_y, u_z on three basic vectors suitably chosen, $\mathbf{a}, \mathbf{b}, \mathbf{c}$, along the x, y, z directions respectively:

$$\mathbf{u} = u_x\mathbf{a} + u_y\mathbf{b} + u_z\mathbf{c}. \tag{9}$$

If possible, these basic vectors are chosen to be of unit length ($|\mathbf{a}| = |\mathbf{b}| = |\mathbf{c}| = 1$) and orthogonal, since this considerably simplifies the work.

In forming the scalar product of $\mathbf{u}$ with

$$\mathbf{v} = v_x\mathbf{a} + v_y\mathbf{b} + v_z\mathbf{c}, \tag{10}$$

it is easily proved, as done in books on vector algebra, that the expressions on the right can be used exactly as polynomials, that is

$$\mathbf{u} \cdot \mathbf{v} = u_x v_x \mathbf{a} \cdot \mathbf{a} + u_x v_y \mathbf{a} \cdot \mathbf{b} + u_x v_z \mathbf{a} \cdot \mathbf{c} + u_y v_x \mathbf{b} \cdot \mathbf{a} + \ldots + u_z v_z \mathbf{c} \cdot \mathbf{c} \tag{11}$$

If the basic vectors are of unit length and orthogonal, this expression can be considerably simplified since it follows from their definition and the results (i) and (ii) given above that

$$\left. \begin{array}{l} \mathbf{a} \cdot \mathbf{a} = \mathbf{b} \cdot \mathbf{b} = \mathbf{c} \cdot \mathbf{c} = 1, \\[2ex] \mathbf{a} \cdot \mathbf{b} = \mathbf{b} \cdot \mathbf{c} = \mathbf{c} \cdot \mathbf{a} = 0, \end{array} \right\} \tag{12}$$

in which case the only terms that survive in (11) are those that contain $\mathbf{a} \cdot \mathbf{a}$, $\mathbf{b} \cdot \mathbf{b}$, and $\mathbf{c} \cdot \mathbf{c}$:

$$\mathbf{u} \cdot \mathbf{v} = u_x v_x + u_y v_y + u_z v_z. \tag{13}$$

The right-hand side of (13) looks exactly of the same form as the expression $k_x r_x + k_y r_y + k_z r_z$ that we found in the exponent of the Bloch functions (8). In fact, on comparing (1) and (9), r_x, r_y, r_z are the components of **r**, so that we are tempted to write the exponent of the Bloch function as a scalar product of two vectors, one of which is **r** and the other is a vector of components k_x, k_y, k_z. Unfortunately, this cannot be done straightaway, since (13) is valid only when the basic vectors are unit and orthogonal whereas, as we have said, the basic vectors that we have to use in a crystal (the primitive vectors) are seldom othogonal and never unit.

It is possible nevertheless to recover the simplicity of (13) even when we deal with non-orthogonal non-unit basic vectors, by using a mathematical artifact called the *reciprocal vectors*. Suppose that we have three non-orthogonal basic vectors **a, b, c**. We shall introduce a second, auxiliary, set of vectors **a*, b*, c*** chosen so cunningly that between the two sets we can mock up the relations (12) that lead to the simple form (13) of the scalar product.† In fact, we shall construct **c*** perpendicular to **a** and **b** (i.e. perpendicular to their plane, which is always possible— see Fig. 54) and such that $\mathbf{c}.\mathbf{c}^* = 1$ and in a similar way two other vectors **a*** and **b***:

$$\mathbf{a}.\mathbf{a}^* = \mathbf{b}.\mathbf{b}^* = \mathbf{c}.\mathbf{c}^* = 1,$$

$$\mathbf{a}.\mathbf{b}^* = \mathbf{b}.\mathbf{c}^* = \mathbf{c}.\mathbf{a}^* = \mathbf{a}^*.\mathbf{b} = \mathbf{b}^*.\mathbf{c} = \mathbf{c}^*.\mathbf{a} = 0. \qquad (14)^*$$

[Notice, in fact, that in (14) just as in (12) the scalar product between two vectors with the same name is unity and between two vectors of different name is nought.] The vectors **a*, b*, c*** are called the *reciprocal vectors* of **a, b, c**.

Before we give an example of the reciprocal vectors let us proceed formally and see how they allow us to obtain a simple expression for the scalar product of two vectors **a** and **b**. We write **u** just as in (9),

---

† So far in this book the asterisk has been used to denote the complex conjugate, and the reader should not confuse this meaning with the new one: when the asterisk is used on vectors, **a**, or on their components, a_x, etc., or on vectorial symbols such as [*hkl*] (see § 4), it always means that the corresponding quantities are taken in the reciprocal lattice. In every other case, the asterisk denotes complex conjugation.

$$\mathbf{u} = u_x\mathbf{a} + u_y\mathbf{b} + u_z\mathbf{c}, \tag{15}$$

but the next step in the trick is to express $\mathbf{v}$ in terms of its components on the reciprocal vectors $\mathbf{a}^*$, $\mathbf{b}^*$, $\mathbf{c}^*$ (if these are known there is no trouble in using them as basic vectors to decompose any vectors we may want in terms of them). We shall call these *reciprocal components v_x^*, v_y^*, v_z^**. That is

$$\mathbf{v} = v_x^*\mathbf{a}^* + v_y^*\mathbf{b}^* + v_z^*\mathbf{c}^*. \tag{16}$$

We form the scalar product of (15) and (16) in the same manner as that of (9) and (10) which led to (11):

$$\mathbf{u}\cdot\mathbf{v} = u_x v_x^*\mathbf{a}.\mathbf{a}^* + u_x v_y^*\mathbf{a}.\mathbf{b}^* + u_x v_z^*\mathbf{a}.\mathbf{c}^* + u_y v_x^*\mathbf{b}.\mathbf{a}^* + \ldots + u_z v_z^*\mathbf{c}.\mathbf{c}^*. \tag{17}$$

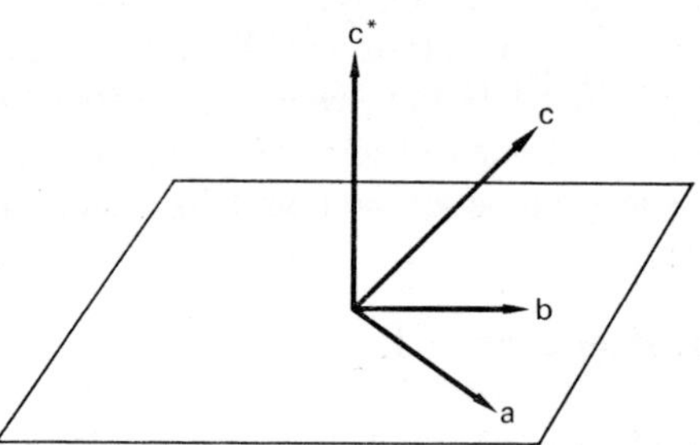

FIG. 54. A reciprocal vector. $\mathbf{c}^*$ is perpendicular to $\mathbf{a}$ and $\mathbf{b}$ and its length is adjusted so that $\mathbf{c}^*\cdot\mathbf{c} = 1$.

On applying conditions (14), the only terms that survive in the right-hand side are those that contain vectors of the same name, exactly as before, so that

$$\mathbf{u}\cdot\mathbf{v} = u_x v_x^* + u_y v_y^* + u_z v_z^*, \tag{18}$$

which is of the same form as (13). Briefly, it is enough to remember that we can work with non-orthogonal basic vectors under the same rules as with orthogonal ones as long as we always mix direct with reciprocal components.

We give a two-dimensional example of the reciprocal vectors in Fig. 55, where, for simplicity, we assume that the basic vectors **a** and **b** are of unit length: $|\mathbf{a}| = |\mathbf{b}| = 1$. The direction of **a*** and **b*** can be chosen at once, since they must be orthogonal to **b** and **a** respectively. We have chosen the length of **b*** such that $|\mathbf{b}^*| \cos \theta = |\mathbf{b}|$ and therefore $\mathbf{b}^* \cdot \mathbf{b} = |\mathbf{b}^*| \cos \theta |\mathbf{b}| = |\mathbf{b}|^2 = 1$ and in the same manner for **a***. It can immediately be seen that conditions (14) are satisfied, so that **a***, **b*** are the reciprocal vectors of **a**, **b**. We also show in Fig. 55 two vectors **u** and **v** and how they must be decomposed as required by (15) and (16).

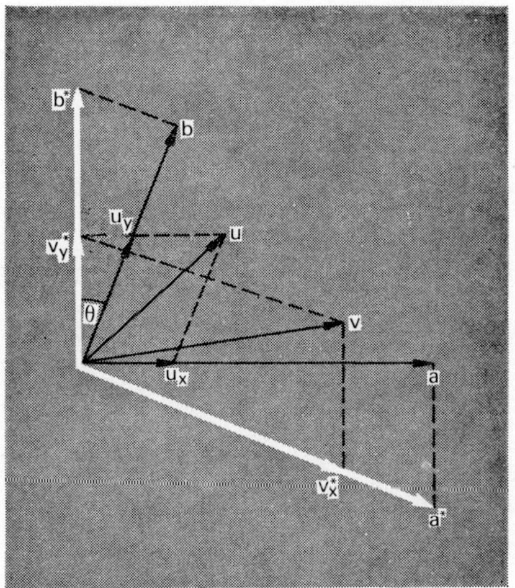

Fig. 55. Reciprocal vectors and reciprocal components. Notice that the vector **v** can also be expressed in terms of direct components $\mathbf{v}_x$ and $\mathbf{v}_y$ (not shown). This is why it is only the components of a vector, and not the vector itself, that are starred in the reciprocal lattice.

The reciprocal vectors trick may look just too twisted. Certainly, if we had to find the scalar product of only one pair of vectors, we would probably compute it more quickly by brute force. But if we have a crystal and we have to compute the scalar product of several hundred pairs of vectors defined in it, the extra work of finding once for all the reciprocal vectors to the crystal axes pays off very handsomely.

4. Reciprocal lattice

In the same manner that by periodic repetition the vectors **a**, **b**, **c** span the crystal (or direct) lattice, the periodic repetition of **a***, **b***, **c*** generates a new lattice, which is called the *reciprocal lattice*. As an example, we show in Fig. 56 how the vectors **a**, **b** and **a***, **b*** defined in Fig. 55 generate the direct and reciprocal lattices respectively. It can be seen that the introduction of such lattices is useful in order to determine quickly the components of vectors, in particular those that have integral components.

The technique given in the last section requires expressing one vector of each pair of vectors in the direct lattice and the other in the reciprocal.

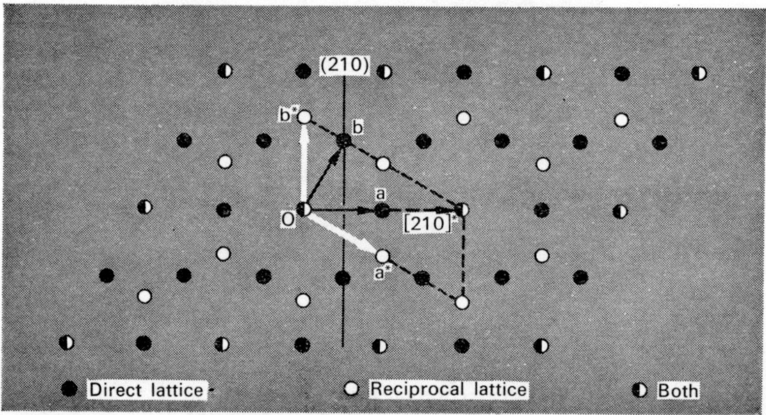

FIG. 56. An hexagonal close-packed layer and its reciprocal lattice. The points of the direct lattice are at the corners and centres of hexagons. The reciprocal lattice is also hexagonal close packed. The plane whose trace is the solid vertical line shown has Miller indices (210) since it cuts the axes **a**, **b**, **c** at $\frac{1}{2}$, 1, and ∞ respectively. (The Miller indices are the inverses of these numbers.) The reciprocal lattice vector [210]* has components 2, 1, and 0 along **a***, **b***, and **c*** respectively. (**c** and **c*** are perpendicular to the drawing.)

This may appear to be chaotic but it can be done very neatly in practice, since in most problems one class of vectors will more naturally be expressed in the direct lattice. In the exponent of (8), for instance, the vector of components r_x, r_y, r_z is the position vector $\mathbf{r}$ of a point of the crystal, as discussed, and it is natural to represent it in the direct lattice. We shall see later that the other vector, of components k_x, k_y, k_z denotes the direction of a wave that propagates along the crystal and, since this is more loosely connected to the crystal, it is not inconvenient to represent it in the reciprocal lattice. We can thus separate all the vectors that appear in a problem in two physically distinct classes (such as position and propagation vectors in the example above), which permits us to separate entirely the representations of the two lattices: we can have, say, one drawing of the direct lattice in which all position vectors will be plotted and a second, separate, drawing of the reciprocal lattice in which all propagation vectors are represented. If this is done, which is most often the case, it is essential that the physical interpretation of the vectors in the reciprocal lattice be made only after the latter is properly oriented with respect to the real lattice: if, for instance, the two lattices of Fig. 56 are traced on two separate sheets and the reciprocal-lattice tracing is inspected on its own, there is nothing in it to indicate the direction of propagation of the vector [210]* with respect to the crystal, although a rule will be given later to obtain this direction. (Beginners often make the mistake of reading vectors in the reciprocal lattice as if their directions were straightaway directions in the direct lattice.)

Since, as we said, one normally deals with the reciprocal lattice on its own, it would be very tiresome to have to orient it with respect to the direct one every time the true direction of a vector is required. Fortunately, the following result, which is proved in Exercise 1 below, allows us to dispense with this step.

Denote, in the standard crystallographic notation as so far used, planes with Miller indices (hkl) and directions with vectors [mnp], where m, n, p, are the components in terms of the basic vectors of the vector that specifies the desired direction. When such symbols refer to the reciprocal lattice we mark them with an asterisk. The result in question can now be enunciated: the direction [hkl]* in the reciprocal space is perpendicular to the planes (hkl) in direct space. Since, as we have indicated, we shall use reciprocal space to denote directions of propagation of waves in the (real) crystal, the physical significance of a direction as read from a reciprocal lattice picture can immediately be obtained even without orienting it with respect to the direct lattice. For instance, if in Fig. 56 we had nothing more than the reciprocal lattice, we would still read correctly the vector [210]* as a direction perpendicular to the real (210) plane, which is in fact apparent from the figure.

To complete the description of the reciprocal lattice we now discuss the moduli of its basic vectors. We know that **c*** is perpendicular to **a** and **b** and hence to their plane (see Fig. 57). The modulus of **c*** is obtained from the condition **c*** . **c** = 1, that is

$$|\mathbf{c}^*|\,|\mathbf{c}|\cos\theta = 1. \tag{19}$$

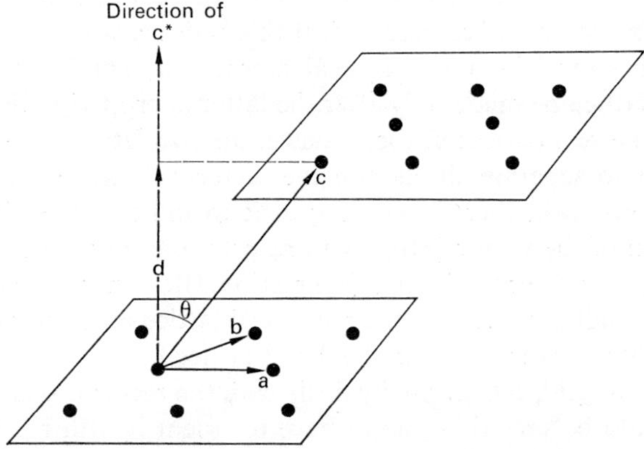

FIG. 57. Relation between **c*** and the interplanar spacing. **a** and **b** determine a crystal layer or plane. **c** gives the distance in Å between two adjacent layers, along the direction shown. *d* is the interplanar spacing in Å.

(We must remember that, as discussed in § 1, the basic vectors in a crystal are not of unit length.) From the figure, $|\mathbf{c}|\cos\theta = d$, the interplanar spacing. Hence, from (19)

$$|\mathbf{c}^*| = d^{-1}. \tag{20}$$

That is, the length of each basic vector of the reciprocal lattice is the reciprocal of the interplanar distance of the stack of planes to which it is perpendicular.

It can be proved in general that the modulus of the vector $[hkl]^*$ in reciprocal space is the reciprocal of the interplanar distance of the (hkl) planes of direct space (see Exercise 2).

In summary: the basic vectors of a crystal are neither orthogonal nor unit: the introduction of the reciprocal vectors allows us to work as if we had a system of orthogonal unit vectors.

Exercise 1

Prove that the vector $[hkl]^*$ is perpendicular to the plane (hkl).

Method. The plane (hkl) is represented in Fig. 58. Prove that $[hkl]^*$ is perpendicular to the two vectors $\mathbf{m}$ and $\mathbf{n}$ of (hkl). Observe that $\mathbf{m} = -(1/h)\mathbf{a}+(1/k)\mathbf{b}$ and that $[hkl]^* = h\mathbf{a}^*+k\mathbf{b}^*+l\mathbf{c}^*$. Verify that the scalar product $\mathbf{m} \cdot [hkl]^* = -\mathbf{a} \cdot \mathbf{a}^* + \mathbf{b} \cdot \mathbf{b}^* = 0$. Show in the same manner that $\mathbf{n} \cdot [hkl]^* = 0$ and the theorem is proved.

Exercise 2

Call $|[hkl]^*|$ the modulus of $[hkl]^*$ and $d(hkl)$ the interplanar spacing of the planes (hkl). Prove that

$$d(hkl) = |[hkl]^*|^{-1}. \qquad *$$

Method. We can always consider in Fig. 58 that O has been taken on one of the (hkl) planes, so that the distance OP from O to the plane (hkl) is $d(hkl)$. If $\mathbf{v}$ is *any* vector from O to the plane and $\mathbf{u}$ is a unit vector perpendicular to the plane, $OP = d(hkl) = \mathbf{v} \cdot \mathbf{u}$. In fact, $\mathbf{v} \cdot \mathbf{u}$ is the projection of $\mathbf{v}$ along $\mathbf{u}$ (which is OP) times the modulus of $\mathbf{u}$ (which is unity). From Exercise 1 we can take $\mathbf{u} = [hkl]^*/|[hkl]^*|$. Also, take $\mathbf{v} = [(1/h)00]$. Therefore

$$OP = d(hkl) = \left[\frac{1}{h}\,00\right] \cdot \frac{[hkl]^*}{|[hkl]^*|} = \frac{1}{|[hkl]^*|}.$$

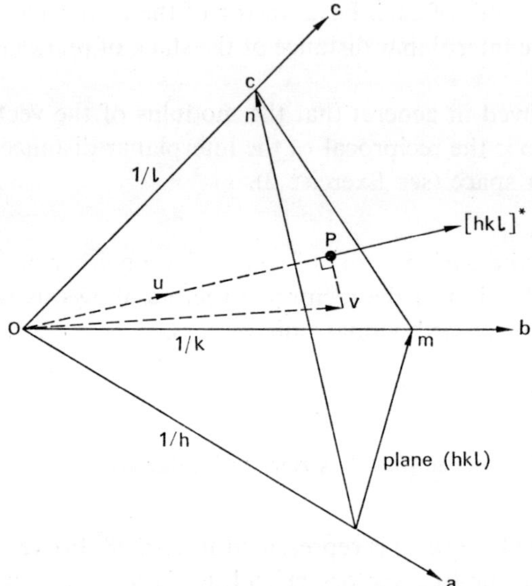

FIG. 58. [*hkl*]* is perpendicular to (*hkl*).
P is the intersection of the plane (*hkl*) with
the perpendicular to it from *O*.

Exercise 3

(This exercise and the following one require a knowledge of vector products and mixed triple products.)

Verify that if **a**, **b**, **c** are direct vectors, their reciprocal vectors are given by

$$\mathbf{a}^* = \frac{\mathbf{b} \times \mathbf{c}}{\mathbf{a} \cdot \mathbf{b} \times \mathbf{c}} \qquad (21)^*$$

and similar equations obtained by cyclic permutation (i.e. the changes $\mathbf{a}^* \to \mathbf{b}^* \to \mathbf{c}^* \to \mathbf{a}^*$, $\mathbf{a} \to \mathbf{b} \to \mathbf{c} \to \mathbf{a}$).

Method. Obviously, $\mathbf{a} \cdot \mathbf{a}^* = 1$. Also, $\mathbf{b} \cdot \mathbf{a}^* = 0$ because the numerator $\mathbf{b} \cdot \mathbf{b} \times \mathbf{c}$ contains a mixed triple product two vectors of which are equal. (Such triple products always vanish.)

Exercise 4

Show that if Ω and Ω^* are the volumes spanned by the vectors $\mathbf{a}, \mathbf{b}, \mathbf{c}$ and $\mathbf{a}^*, \mathbf{b}^*, \mathbf{c}^*$ respectively, $\Omega^* = \Omega^{-1}$.

Method. It follows from vector methods that

$$\Omega = \mathbf{a}.\mathbf{b}\times\mathbf{c}, \qquad \Omega^* = \mathbf{a}^*.\mathbf{b}^*\times\mathbf{c}^*.$$

Substitute in Ω^* the values of $\mathbf{a}^*, \mathbf{b}^*, \mathbf{c}^*$ given by (21):

$$\Omega^* = \Omega^{-3}\{(\mathbf{b}\times\mathbf{c}).[(\mathbf{c}\times\mathbf{a})\times(\mathbf{a}\times\mathbf{b})]\}.$$

Use the formula from vector algebra,

$$\mathbf{m}\times(\mathbf{n}\times\mathbf{p}) = (\mathbf{m}.\mathbf{p})\mathbf{n}-(\mathbf{m}.\mathbf{n})\mathbf{p},$$

to prove that the square bracket on the right-hand side is equal to $\Omega\mathbf{a}$. (In doing so remember that $\mathbf{c}\times\mathbf{a}.\mathbf{a} = 0$.)

5. The Bloch functions in vector notation

Let us consider again the Bloch functions (8):

$$\varphi^j_{k_x k_y k_z}(xyz) = \exp[2\pi i(k_x r_x + k_y r_y + k_z r_z)]u^j_{k_x k_y k_z}(xyz). \tag{22}$$

As in (1), we have

$$\mathbf{r} = r_x\mathbf{a}+r_y\mathbf{b}+r_z\mathbf{c}, \tag{23}$$

so that the quantities r_x, r_y, r_z in the exponent of (22) are the components of a vector in the direct lattice. Therefore, for this exponent to represent a scalar product the components of $\mathbf{r}$ must be multiplied by the reciprocal-lattice components of another vector. It appears at first that k_x, k_y, k_z must be components of a vector in the direct lattice, on account of the suffices that they have, but if we look again at (4), (5), and (6) we see that this is not necessarily so. All that we state in those equations is that k_x, k_y, k_z are some numbers defined by (7); the suffices used do not have any vector meaning: they merely indicate that k_x, say, is the number that corresponds to $\varphi^j(x)$. We could just as well have labelled these numbers as k_X, k_Y, k_Z or k_ξ, k_η, k_ζ or in any other way as long as the notation indicates somehow which are the Bloch functions to which the k numbers pertain. We shall in fact relabel them as k_x^*, k_y^*, k_z^* and shall accordingly define a vector

$$\mathbf{k} = k_x^*\mathbf{a}^* + k_y^*\mathbf{b}^* + k_z^*\mathbf{c}^* \tag{24}$$

with $k_x^* = \kappa_x/N_x$, etc. [see (7)].

(All that we have done is this: we have taken the basic vectors $\mathbf{a}^*, \mathbf{b}^*, \mathbf{c}^*$ and we have formed a vector $\mathbf{k}$ the component of which along $\mathbf{a}^*$ is the number k_x given by (7), etc. Since k_x, k_y, k_z are now the components of this vector along the reciprocal basic vectors, we relabel them as k_x^*, k_y^*, k_z^*.)

It follows from (23) and (24) that the (relabelled) exponent in (22) is

$$k_x^*r_x + k_y^*r_y + k_z^*r_z = \mathbf{k}.\mathbf{r},$$

so that (22) can be rewritten in the compact vector form

$$\varphi_{\mathbf{k}}^j(\mathbf{r}) = \exp(2\pi i\mathbf{k}.\mathbf{r})u_{\mathbf{k}}^j(\mathbf{r}), \tag{25}$$

where $u_{\mathbf{k}}^j(\mathbf{r})$ is periodic in the direct lattice: $u_{\mathbf{k}}^j(\mathbf{r}) = u_{\mathbf{k}}^j(\mathbf{r}+\mathbf{R})$, $\mathbf{R}$ being a lattice translation vector defined in (2).

Since we shall *always* define the components of the $\mathbf{k}$ vectors in the reciprocal lattice, we shall write k_x, k_y, k_z for k_x^*, k_y^*, k_z^* in order to simplify the notation.

We can easily check that, just as in one dimension, the Bloch functions (25) are eigenfunctions of the translation operators $\mathbf{R}$. In fact, $\mathbf{R}$ is the translation operator that transforms the position vector $\mathbf{r}$ into $\mathbf{r}+\mathbf{R}$, $\mathbf{R}$ being the translation vector that corresponds to the operator $\mathbf{R}$. Hence

$$\mathbf{R}\varphi_{\mathbf{k}}^j(\mathbf{r}) = \exp[2\pi i\mathbf{k}.(\mathbf{r}+\mathbf{R})]u_{\mathbf{k}}^j(\mathbf{r}+\mathbf{R})$$

$$= \exp(2\pi i\mathbf{k}.\mathbf{R})\varphi_{\mathbf{k}}^j(\mathbf{r}). \tag{26}$$

Therefore, just as in one dimension, the label $\mathbf{k}$ of the Bloch function denotes the eigenvalue of $\mathbf{R}$ to which it belongs.

6. The wave vector

In order to find the physical meaning of the vector $\mathbf{k}$ in (25) we shall compare the Bloch function with the free-electron eigenfunction into which it goes in the weak-field limit. This was obtained in § 2.6:

$$\psi(xyz) = \exp[2\pi i(k_x x + k_y y + k_z z)].\tag{27}$$

[We now dispense with the normalization factor used in (**2.27**).] For free electrons, the basic vectors $\mathbf{a}, \mathbf{b}, \mathbf{c}$ can always be taken as unit and orthogonal, since there is no crystal lattice to worry about. Therefore, in this case, $x = x/a = r_x$, etc., and the exponent in (27) is straightaway a scalar product:

$$\psi(xyz) = \exp(2\pi i \mathbf{k}.\mathbf{r}),\tag{28}$$

where $\mathbf{r}$ is given by (23) and $\mathbf{k} = k_x\mathbf{a} + k_y\mathbf{b} + k_z\mathbf{c}$.

The meaning of $\mathbf{k}$ in this case follows from the fact that the momentum vector $\mathbf{p}$ is

$$\mathbf{p} = p_x\mathbf{a} + p_y\mathbf{b} + p_z\mathbf{c} = h(k_x\mathbf{a} + k_y\mathbf{b} + k_z\mathbf{c}) = h\mathbf{k},\tag{29}$$

since $p_x = hk_x$, etc. From (29) the vector $\mathbf{k}$ is parallel to the momentum, so that it is a vector in the direction of propagation of the free electron. It can also be seen that $|\mathbf{k}| = \lambda^{-1}$, as was the case before for the wave number k. This vector $\mathbf{k}$ along the direction of propagation of a free particle and of modulus equal to the wave number, is called the *wave vector*.

If we now go back to the Bloch function (25) we see that in the limit of weak fields $u_{\mathbf{k}}^j(\mathbf{r})$ must become a constant so that the Bloch function tends correctly to the free-electron wave function (28). Correspondingly, the vector $\mathbf{k}$ in the Bloch function must have the asymptotic meaning of a wave vector and, as anticipated in § 4, it indicates the direction of propagation of the Bloch wave. (The reader should avoid thinking that this coincides with the direction of propagation of the particle. This is the case only for a free electron. See § 16.)

It should be noticed that the representation of $\mathbf{k}$ in the reciprocal space is particularly appropriate since, from (20) the basic vectors in this space are measured in units of $Å^{-1}$, just as $\mathbf{k}$ is (since its modulus is λ^{-1}).

7. Symmetry in the reciprocal space

In the same manner that the energy is a function of **k** in the one-dimensional lattice and has to be plotted against the **k** axis to give the $E(k)$ curves, it is in three dimensions a function of the **k** vector and, since the values of the latter are given in reciprocal space, the energy has to be represented by means of surfaces in this space. The correspondence with the one-dimensional case is even closer, since we can show that the **k** space that we defined in one dimension is exactly the reciprocal space of the linear chain lattice. The latter is described by a single primitive vector **a**, the modulus of which (that is the lattice spacing) is a, whence the reciprocal lattice of the linear chain is given by a vector **a*** of modulus $|\mathbf{a}^*| = a^{-1}$. In fact, the periodic lattice in k space found in § 3.10 was given by the vectors K_m of modulus ma^{-1}, which can be written as $\mathbf{K}_m = m\mathbf{a}^*$ in the present notation.

The symmetry conditions (3.43) and (3.44) for the energy as a function of k, $E(k) = E(-k)$, $E(k) = E(k+K_m)$ can be extended to three dimensions. We shall consider the periodicity condition first. If we rewrite it in our present notation, it takes the form $E(\mathbf{k}) = E(\mathbf{k}+m\mathbf{a}^*)$, in accordance with the identification just made. In three dimensions, we must allow for the two other basic periodicities

$$E(\mathbf{k}) = E(\mathbf{k}+n\mathbf{b}^*), \; E(\mathbf{k}) = E(\mathbf{k}+p\mathbf{c}^*),$$

and the three conditions can be written in the single equation

$$E(\mathbf{k}) = E(\mathbf{k}+m\mathbf{a}^*+n\mathbf{b}^*+p\mathbf{c}^*). \tag{30}$$

We shall define

$$\mathbf{K}_{mnp} \equiv m\mathbf{a}^*+n\mathbf{b}^*+p\mathbf{c}^*, \quad m,n,p, \text{ integers}, \tag{31}$$

which is called a *vector of the reciprocal lattice*. This is a vector with integral components and it is the counterpart of the lattice translation vectors $\mathbf{R}_{mnp}$ of the direct lattice [see (2)]. Just as for these vectors, we shall write $\mathbf{K}_{mnp} = \mathbf{K}$ unless more detail is necessary. In this notation (30) becomes

$$E(\mathbf{k}) = E(\mathbf{k}+\mathbf{K}). \tag{32}$$

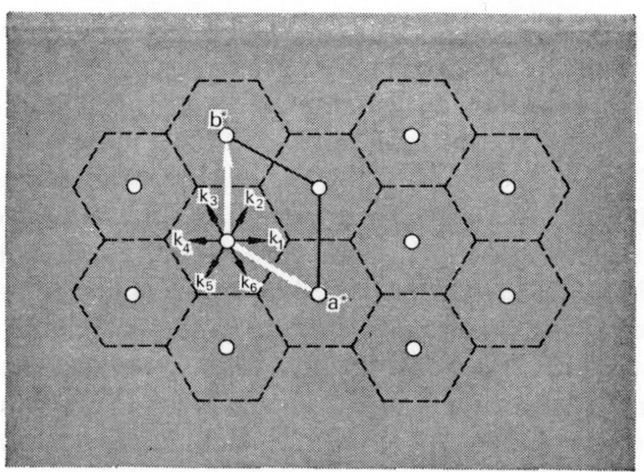

FIG. 59. The reciprocal lattice of a hexagonal close packed layer (cf. Fig. 56). **a*** and **b*** are the primitive vectors of the reciprocal lattice and the parallelogram that they span is the primitive cell in the reciprocal lattice. $\mathbf{k}_1, \mathbf{k}_2, \ldots, \mathbf{k}_6$ form a star. The hexagonal cell, which has the same area as the primitive cell, is a centred primitive cell or Brillouin zone (see § 9). As should be the case for a primitive cell, the whole lattice is covered by periodic repetition of the hexagonal cell by the vectors of the reciprocal lattice.

To appreciate the importance of this relation, let us consider on its own the reciprocal lattice of the hexagonal close packed layer which we derived in Fig. 56. We can see at once from Fig. 59 that it is itself a hexagonal close packed layer, but this is not a very significant fact (in most cases the reciprocal and direct lattices are of a different nature). The important point is that, on account of (32) it is enough to represent $E(\mathbf{k})$ in the primitive cell of the reciprocal lattice: it is then determined over the whole of the reciprocal lattice by translational symmetry. This primitive cell is called the *first Brillouin zone* (see § 9).

Let us now consider what happens in three dimensions with the relation $E(k) = E(-k)$, which, as we saw in § 3.9, is simply the result of the fact that the direct lattice has a centre of inversion besides translational symmetry. We shall first look again in a slightly different form at the work of §§ 3.8 and 3.9.

We must first observe that the direct and reciprocal lattices are indissolubly tied up. Although we may draw them separately, this is purely a matter of convenience: the real situation is as in Fig. 56. This means that if we effect a transformation in the direct lattice we must at the same time transform the reciprocal one. Take, for example, the one-dimensional case where both lattices are defined along the same axis (since we have the single condition $\mathbf{a} \cdot \mathbf{a}^* = 1$ or, more intuitively, since the only direction of propagation that can be represented by means of the reciprocal lattice must be along $\mathbf{a}$ itself). If we invert the direct lattice we must also invert the reciprocal one (Fig. 60), i.e. the inversion of the real lattice induces the transformation of $\mathbf{k}$ into $-\mathbf{k}$ in the reciprocal one. Since the inversion is a symmetry operation, it cannot affect the energy, whence $E(\mathbf{k}) = E(-\mathbf{k})$, as we saw in § 3.9.

In one dimension i is the only symmetry operation apart from the translations, but in three dimensions we may have many more operations and we can use intuitively the above argument in order to find the symmetry properties of the energy in the reciprocal space. (A more rigorous justification requires the use of group theory.) In fact, we see in Fig. 56 that the direct lattice has a sixfold axis of rotation which, of course, also

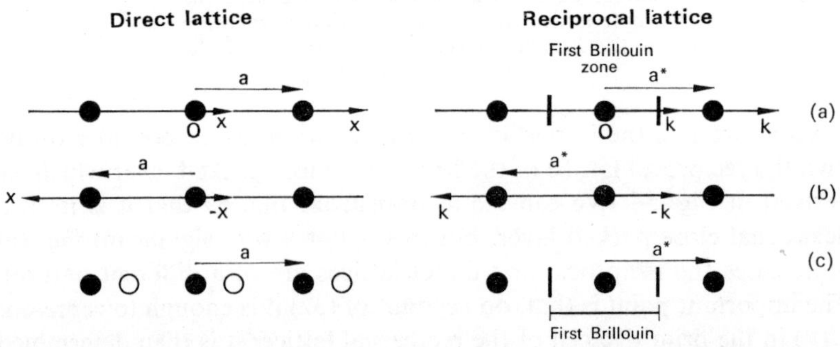

FIG. 60. Inversion in the reciprocal lattice. In (a) $\mathbf{a}^*$ has been chosen so that $\mathbf{a} \cdot \mathbf{a}^* = 1$, where $\mathbf{a}$ is the direct lattice primitive vector chosen on the left. (A suitable scale is assumed for the reciprocal lattice so that $|\mathbf{a}^*| = |\mathbf{a}|^{-1}$.) In (b) the primitive vector of the direct lattice has been inverted and the same must be done with $\mathbf{a}^*$: otherwise $\mathbf{a} \cdot \mathbf{a}^*$ would be -1. On inversion $\mathbf{x} \to -\mathbf{x}$ in the direct lattice and $\mathbf{k} \to -\mathbf{k}$ in the reciprocal one, which was proved in § 3.8 in a different way. In (c) it is shown that the lattice with basis on the left gives exactly the same reciprocal lattice as before.

appears in the reciprocal one (see Fig. 59). Clearly, from Fig. 56, a sixfold rotation of the direct lattice drags with it the reciprocal one, and the six vectors $\mathbf{k}_1, \mathbf{k}_2, \ldots, \mathbf{k}_6$ of Fig. 59 are transformed one into another by this operation. Just as in the one-dimensional case, the energy cannot be altered by such a symmetry operation, so that $E(\mathbf{k}_1) = E(\mathbf{k}_2) = \ldots = E(\mathbf{k}_6)$.

In general, a set of $\mathbf{k}$ vectors that are transformed one into the other by a symmetry operation of the reciprocal lattice is called a *star*: all the vectors of a star correspond to the same energy. In the one-dimensional case we had only one type of star, which contains $\mathbf{k}$ and $-\mathbf{k}$ (except of course when $\mathbf{k} = 0$) which led to the corresponding degeneracy of the energy. We shall now have more symmetry conditions, which can be stated by saying that whenever $\mathbf{k}$ and $\mathbf{k}'$ belong to the same star $E(\mathbf{k}) = E(\mathbf{k}')$.

It cannot be overemphasized that the symmetry of the reciprocal lattice is a vicarious one: it is purely the result of such symmetry operations as are permitted in the direct lattice. This remark is important because, as mentioned in § 3.21, the reciprocal lattice is blind to the contents of the primitive cell. Whereas the nature of the contents of the primitive cell may destroy symmetry operations of the direct lattice, it does not have any influence on the geometry of the reciprocal lattice, which when considered on its own will appear to possess more symmetry than really allowed. Thus, the lattice with basis in Fig. 60c gives a Brillouin zone that differs in no way from those shown in (a) and (b). However, the lattice with basis (c) does not admit of the inversion as a symmetry operation, whereas the Brillouin zone would appear to possess this symmetry. We shall see in § 8 that inversion in the reciprocal lattice happens to have a special status, but the above example should illustrate the fact that the study of the Brillouin zone on its own can lead into error since, as a purely geometrical figure, the Brillouin zone may possess symmetry operations forbidden in the crystal lattice.

8. $E(\mathbf{k}) = E(-\mathbf{k})$: complex conjugation

In accordance with the discussion so far, we would expect the relation $E(\mathbf{k}) = E(-\mathbf{k})$ to hold only when the direct (and hence the reciprocal) lattice admits of a centre of inversion. A rather surprising result is that

F

$$E(\mathbf{k}) = E(-\mathbf{k}) \tag{33}$$

always, whether or not the crystal is centro-symmetric. The reason for this is that there is an operation not yet considered which, although not strictly a symmetry operation, acts very much like one and gives origin to (33). This is the complex conjugation $\mathbf{j}$.

●● 1. Complex conjugation ($\rightarrow$ 19)

Let us call $\mathbf{j}$ the operation of taking the complex conjugate ψ^* of any function ψ: $\mathbf{j}\psi = \psi^*$. We shall show that ψ and $\mathbf{j}\psi$ are degenerate in the same manner that ψ and $\mathbf{R}\psi$ are degenerate if $\mathbf{R}$ is a symmetry operation. In fact, consider the Schrödinger equation

$$\mathbf{H}\psi = E\psi \tag{34}$$

and apply $\mathbf{j}$ on both sides:†

$$\mathbf{H}^*\psi^* = E^*\psi^*. \tag{35}$$

Both the hamiltonian and the energy are real, whence

$$\mathbf{H}\psi^* = E\psi^*, \tag{36}$$

and on comparing (33) with (34) it follows that ψ and ψ^* are degenerate.

We shall prove below the lemma that $\mathbf{j}\varphi_\mathbf{k} = \varphi_{-\mathbf{k}}$, where $\varphi_\mathbf{k}$ is a Bloch function, so that it follows from the result above that $\varphi_\mathbf{k}$ and $\varphi_{-\mathbf{k}}$ are degenerate. Since the first belongs to the eigenvalue $E(\mathbf{k})$ and the second to $E(-\mathbf{k})$, (33) follows.

$$\text{LEMMA. } \mathbf{j}\varphi_\mathbf{k} = \varphi_{-\mathbf{k}}. \tag{37}$$

Proof. Consider the Bloch function

$$\varphi_\mathbf{k} = \exp{(2\pi i \mathbf{k}.\mathbf{r})}\, u_\mathbf{k}^j(\mathbf{r}).$$

---

† Notice that $\mathbf{j}(a\psi) = \mathbf{j}a\mathbf{j}\psi = a^*\psi^*$ whereas a symmetry operator $\mathbf{R}$ cannot modify a constant. Thus $\mathbf{R}(a\psi) = \mathbf{R}a\mathbf{R}\psi = a\mathbf{R}\psi$. Whereas, on account of this behaviour, symmetry operators are linear, $\mathbf{j}$ is *not a linear operator*. (Cf. Problem in § 1.11.)

The label $\mathbf{k}$ denotes the eigenvalue of $\varphi_{\mathbf{k}}$ with respect to a translation [see (26)]:

$$\mathbf{R}\varphi_{\mathbf{k}} = \exp(2\pi i\mathbf{k}.\mathbf{R})\,\varphi_{\mathbf{k}}. \tag{38}$$

Form now $\mathbf{j}\varphi_{\mathbf{k}}$:

$$\mathbf{j}\varphi_{\mathbf{k}} = \exp(-2\pi i\mathbf{k}.\mathbf{r})\,[u_{\mathbf{k}}^{j}(\mathbf{r})]^{*}. \tag{39}$$

In order to find the eigenvalue of $\mathbf{R}$ to which this function belongs, we operate with $\mathbf{R}$ on both sides of (39):

$$\begin{aligned}
\mathbf{R}(\mathbf{j}\varphi_{\mathbf{k}}) &= \exp[-2\pi i\mathbf{k}.(\mathbf{r}+\mathbf{R})]\,[u_{\mathbf{k}}^{j}(\mathbf{r}+\mathbf{R})]^{*} \\
&= \exp(-2\pi i\mathbf{k}.\mathbf{R})\exp(-2\pi i\mathbf{k}.\mathbf{r})\,[u_{\mathbf{k}}^{j}(r)]^{*} \\
&= \exp(-2\pi i\mathbf{k}.\mathbf{R})\,(\mathbf{j}\varphi_{\mathbf{k}}).
\end{aligned} \tag{40}$$

Since $\mathbf{j}\varphi_{\mathbf{k}}$ belongs to an eigenvalue with negative $\mathbf{k}$, the label which corresponds to $\mathbf{j}\varphi_{\mathbf{k}}$ is $\varphi_{-\mathbf{k}}$.

Exercise 1

Prove that $\mathbf{jR} = \mathbf{Rj}$. (Notice that, unlike the inversion, $\mathbf{j}$ commutes with translations.)

Method. Apply $\mathbf{j}$ on both sides of (38) remembering that, on the right-hand side, it operates on both factors, and compare the result with (40).

Note. Exercise 1 is a particular case of Exercise 2.

Exercise 2

Prove that for any symmetry operation $\mathbf{S}$, $\mathbf{jS} = \mathbf{Sj}$.

Method. The operator $\mathbf{S}$ is defined in such a way that, when it acts on an arbitrary function of x, $\varphi(x)$, it transforms the coordinates x into x' (which are specifically related to $\mathbf{S}$):

$$\mathbf{S}\varphi(x) = \varphi(x'). \tag{41}$$

Analogously

$$\mathbf{S}\varphi^*(x) = \varphi^*(x').\tag{42}$$

That is,

$$\mathbf{S}\mathbf{j}\varphi(x) = \varphi^*(x').\tag{43}$$

On applying $\mathbf{j}$ to both sides of (41),

$$\mathbf{j}\mathbf{S}\varphi(x) = \varphi^*(x')\tag{44}$$

and on comparing the left-hand sides of (43) and (44) the result follows.

Note. For spinless particles the operation of complex conjugation coincides with a very important operation which is called *time reversal*. The existence of $\mathbf{j}$ besides determining the inversion symmetry of the energy in $\mathbf{k}$ space even in the absence of a centre of inversion, has important consequences that will be discussed in § 19.

9. The first Brillouin zone

We found it convenient in the one-dimensional work of Chapter **3** to change the basic unit cell of Fig. 32 into the centred one of Fig. 33 so as to display the inversion symmetry of the energy in $\mathbf{k}$ space. We notice a similar situation in Fig. 59: the sixfold symmetry $E(\mathbf{k}_1) = E(\mathbf{k}_2) = E(\mathbf{k}_3) = \ldots = E(\mathbf{k}_6)$ cannot be displayed at all in the primitive cell spanned by the primitive vectors $\mathbf{a}^*$ and $\mathbf{b}^*$. This, in fact, is the primitive cell which is standard in crystallography.

Another type of primitive cell, a *centred primitive cell*, will now be defined which fully reflects the symmetry of the reciprocal lattice: a bundle of translational vectors is first constructed around the origin of reciprocal space by taking the primitive vectors and their resultants and repeating them in accordance with the full rotational symmetry of the lattice (see Fig. 59). The faces of the first centred primitive cell are those perpendicular bisector planes of the bundle so defined that circumscribe the smallest polyhedron round the origin. In the two-dimensional example of Fig. 59 this rule gives the hexagonal cell shown, which now contains all the vectors in any given star and therefore permits the full use of the lattice symmetry.

It can easily be proved that the centred cell has the same area (or in the general case the same volume) as the original one. The reader can do this by splitting the surfaces given in appropriate pieces. It is simpler to argue that both cells cover the lattice by periodic repetition with the same translational vectors, whence it is intuitively clear that they must have the same area.

The first centred primitive cell in reciprocal space which we have just defined is called the *first Brillouin zone*.

We can look at the properties of the Brillouin zone in a slightly different way. We reproduce in Fig. 61 a primitive cell (shaded) in the reciprocal lattice of the hexagonal close packed layer as well as the first centred primitive cell or Brillouin zone. A primitive cell, whether centred or not, can be defined as the set of all points such that (i) they are all translationally inequivalent, and (ii) there are no other points in the plane translationally inequivalent to them. Thus we can take as the definition of a primitive cell that of being a *maximal set of translationally inequivalent points*. Why we adopt such a definition is clear since such a set of points in the reciprocal space is all that we need in order to express the energy as a function of **k**.

Under the terms of this definition, we construct a centred primitive cell as follows. We first draw in Fig. 61 the pair of planes that bisect

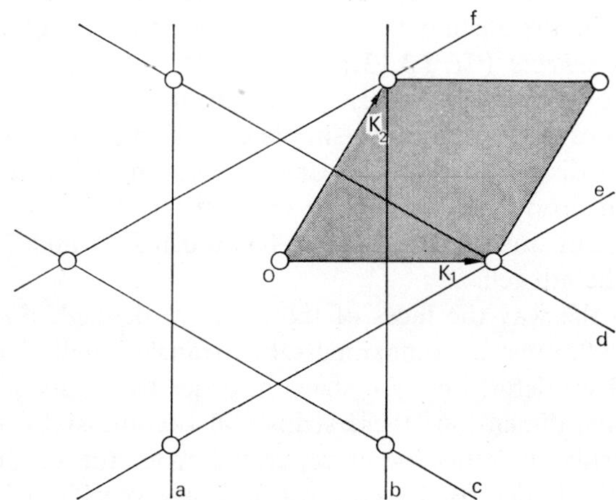

FIG. 61. Construction of the first Brillouin zone.

$\mathbf{K}_1$ and $-\mathbf{K}_1$, labelled a and b in the figure. All points of the plane outside the strip enclosed by a and b must be eliminated, since they are equivalent to points inside the strip by translation with $\mathbf{K}_1$. In the same manner, we must eliminate all points outside the strips enclosed by the pairs of planes c, d and e, f respectively. The figure left, which is the area common to the three strips and limited by the perpendicular bisector planes, is a primitive cell.

10. Brillouin zones of higher order

On account of the periodic condition (32) the energy in a band will be fully determined if it is given in the first Brillouin zone. We can plot each band in this way and therefore give the energy as a multi-valued function in the first Brillouin zone. This is the *reduced scheme* (see § 3.13). On the other hand, as discussed in that section, if we wish to make contact with the free-electron picture, we must display the energy in such a way that the whole of $\mathbf{k}$ space is used. Because of (32) we shall plot each band in a primitive cell (not necessarily the first) and in order to display the $\mathbf{k}$, $-\mathbf{k}$ symmetry (33) these cells must be centred. Just as in § 3.13, therefore, we shall define *nested centred primitive cells in reciprocal space*, called the first, second, etc. Brillouin zones and we shall plot the first band in the first zone, the second band in the second zone, and so on, thereby obtaining the *extended zone scheme*. (Cf. § 3.13.)

We shall now discuss how these nested primitive cells are obtained. First, the idea of *nesting*. This is a simple extension to three dimensions of the concept discussed in § 3.13. Cells are nesting when the $(n+1)$th cell is a polyhedron that contains all the first n cells inside it. (Just as a Chinese nest of boxes.) Thus, the $(n+1)$th cell is bounded inside by the faces of the nth cell.

Now as to the way the faces of the cells are defined. Each cell, as described in § 9, must be a maximal set of translationally inequivalent points. In § 9 we defined certain strips in order to specify such sets of points. In three dimensions these strips will become slabs, defined as follows. Consider a lattice point separated from the origin $\mathbf{O}$ by a vector $\mathbf{K}$ of the lattice (this vector will define a specific neighbour to the origin, first, second, third, etc., in accordance to its distance from it).

There must always be a lattice point separated from **O** by $-\mathbf{K}$ (Fig. 62). Form the planes that bisect perpendicularly **K** and $-\mathbf{K}$. These planes determine a slab such that the points outside the slab (as P' in Fig. 62) are always translationally equivalent to some point inside. The first thing to notice, therefore, is that in order to form the desired primitive cells the *outside* of any such slab must be removed from any cell that is considered, so that no slab can be allowed to remain *inside* any one cell. Rather, the planes of the slabs must always form boundaries of the cells, and inside the cells no such planes (which bisect neighbours to the origin) can exist.

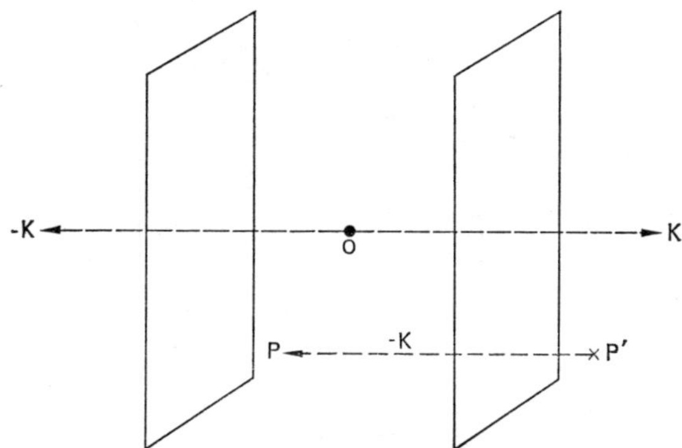

FIG. 62. Faces of the Brillouin zones.

The way in which the successive Brillouin zones are constructed is much simpler than the discussion so far may make appear, and is illustrated for the two-dimensional square lattice in Fig. 63. Firstly, we draw the perpendicular bisector planes to first, second, third, and higher order neighbours from the origin. Secondly, we start from the origin and go away from it until the first set of planes is met. The region inside these planes is the first Brillouin zone. We then start from the surface of the first Brillouin zone and move away from it until we hit the first new bisector planes. This is the second Brillouin zone. In general, we start from the surface of the nth zone and move away from it until bisector planes are met. The non-necessarily connected region of space so defined is the $(n+1)$th Brillouin zone.

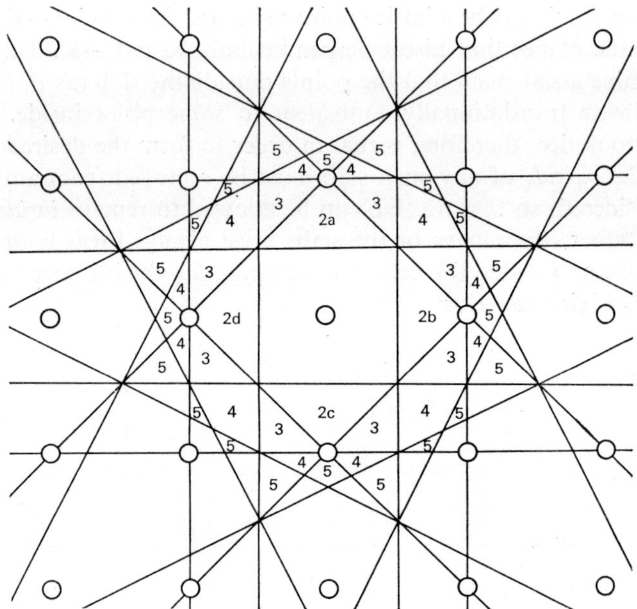

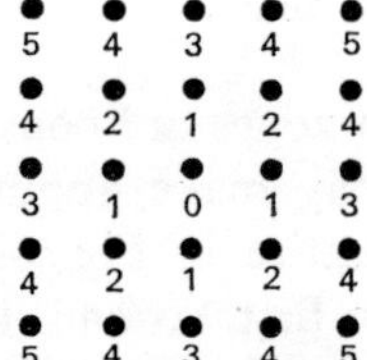

Key to order of neighbours

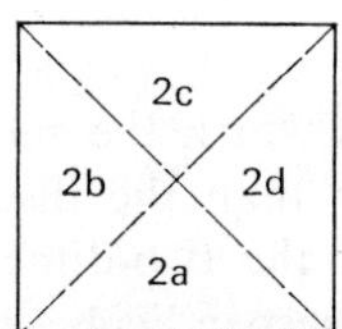

**Folding back the
second zone**

FIG. 63. Brillouin zones for the square lattice. For simplicity, bisector planes to neighbours of order higher than four have not been drawn. The numbers in the main figure identify the first five Brillouin zones and are not directly related to the order of neighbours shown in the key.

The definition given for the successive Brillouin zones ensures that they form maximal sets of translationally inequivalent points. Thus they must have exactly the same volume as the first of such sets or first primitive cell.† They are all, of course, primitive cells. Since, on account of the periodicity condition (32) any part of **k** space can always be translated into an equivalent position by means of a vector **K** of the reciprocal lattice, it is possible to *fold back* the various pieces that make up a higher zone onto the first one. As an example of this process, the inset in Fig. 63 shows how to do this for the second zone. The reader should identify the **k** vectors used in the necessary translations and should also carry out the same operation for the third zone.

PROBLEM. Form the first three Brillouin zones for the reciprocal lattice shown in Fig. 59 and show how they can be folded back on to the first zone.

11. Number of states in the Brillouin zone

We shall now count the number of states in a Brillouin zone. Consider a direct lattice in which we have N_x, N_y, and N_z cells (primitive or unit) respectively along the x,y,z directions. The total number of cells, N, in this lattice will be $N = N_x N_y N_z$. On the other hand, owing to the quantization conditions (7), there are only N_x, N_y, N_z values of **k** allowed in each of the basic vectors **a***, **b***, **c*** of the reciprocal lattice, so that the total number of allowed values of **k** in the cell of the reciprocal lattice is $N_x N_y N_z = N$. (We illustrate this situation in a two-dimensional example in Fig. 64.) It follows that the total number states, that is of permitted values of **k**, in one Brillouin zone is equal to the total number of cells in the crystal, which means, since we can put two electrons per state, that *the Brillouin zone can accommodate two electrons per unit or primitive cell*. (If the lattice is simple, that is, without a basis, this is the same as two electrons per atom. See § 3.21.)

A Brillouin zone must be visualized as containing a uniform mesh of allowed values of **k**, as in the cell shown in Fig. 64. In practice, for N of the order of the Avogadro number, this mesh will be so finely

---

† For a rigorous proof of this result see L. Brillouin, *Wave Propagation in Periodic Structures*, McGraw-Hill, New York, 1946.

F*

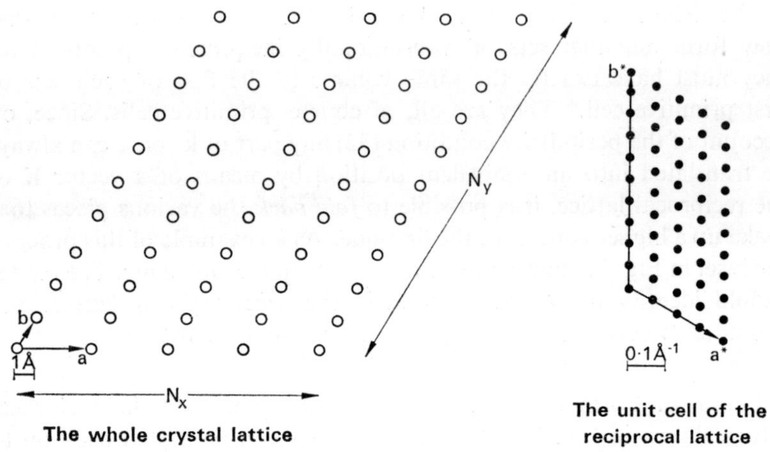

FIG. 64. Number of states in the Brillouin zone.

grained that it can be thought of as practically continuous (*quasi-continuous*). Of course, since the Brillouin zones of higher order are also primitive cells and therefore of the same volume as the first one, they contain exactly the same number of states. Also, since the distribution of permitted **k** values is uniform throughout the Brillouin zone, the number of states in any given portion of it is proportional to its volume (or area in two dimensions).

PROBLEM. It can be seen from the scale in Fig. 64 that $|\mathbf{a}| = 4\,\text{Å}$, $|\mathbf{b}| = 2\,\text{Å}$. The angle between these two vectors is 60°. Verify that the shape and dimensions of the reciprocal-lattice unit cell shown are correct.

12. Conditions for the faces of the Brillouin zones

We shall now find the general conditions satisfied by **k** vectors that end at the surface of a Brillouin zone. The point v in Fig. 65 represents the vth neighbour to the origin in the reciprocal lattice space and the vertical lines shown (bisectors of the line from the origin to v) are faces of a Brillouin zone. It follows at once from the figure that any vector **k** that ends at the face of a Brillouin zone must satisfy this condition: given such a vector, there is always another one $\mathbf{k}'$ of the same length that differs from it by a vector of the reciprocal lattice [see (31)]:

$$\left. \begin{aligned} |\mathbf{k}| &= |\mathbf{k}'|, \\[6pt] \mathbf{k}' &= \mathbf{k}+\mathbf{K}. \end{aligned} \right\} \quad (45)$$

These are, in fact, the von Laue conditions for X-ray diffraction, such as are used in the so-called Ewald construction. That we again make a tie up with X-ray diffraction theory (see § 3.17) should not be surprising since we can now handle three-dimensional problems: the Bragg conditions are essentially two-dimensional and are superseded in the more general case by those of von Laue.

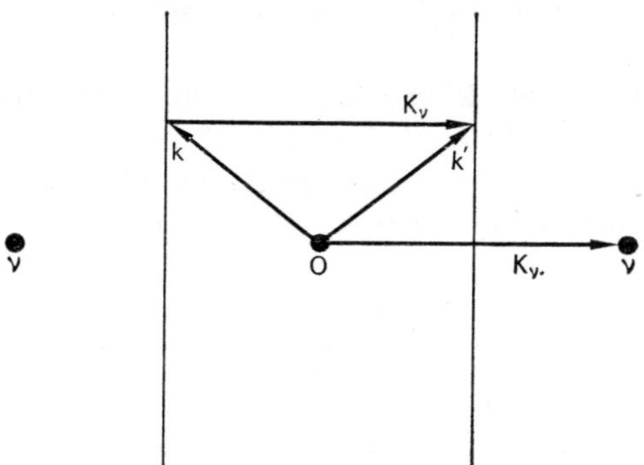

FIG. 65. Conditions for the faces of the Brillouin zones.

We can interpret the conditions (45) as follows. Consider in Fig. 66 a wave of momentum $\mathbf{\varkappa}$ and assume that, on keeping the direction of this vector, its wave number is increased until it reaches the value $\mathbf{k}$ at which it touches the face of the Brillouin zone. Since $\mathbf{k}'$ in Fig. 66 differs from $\mathbf{k}$ by a vector of the reciprocal lattice, it can be substituted for $\mathbf{k}$, which can therefore be considered to have been *reflected by the lattice*. That the representative vector of our wave has to be changed from $\mathbf{k}$ into $\mathbf{k'}$ is clear if we increment further $\mathbf{k}$ into $\boldsymbol{\lambda}$, as in the figure. This vector, being outside the first unit cell in $\mathbf{k}$ space, has to be replaced by the equivalent vector $\boldsymbol{\lambda'}$, so that the increment of the vector from $\mathbf{k}$ to $\boldsymbol{\lambda}$ has to be represented by that from $\mathbf{k'}$ to $\boldsymbol{\lambda'}$.

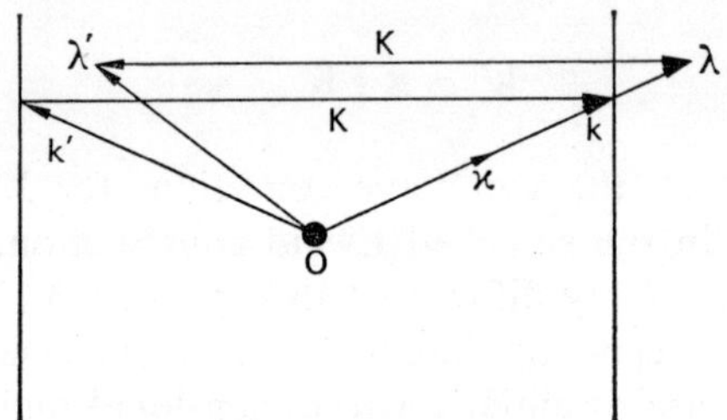

FIG. 66. The vertical lines are faces of the Brillouin zone separated by the vector $\mathbf{K}$ of the reciprocal lattice.

The reflection of $\mathbf{k}$ into $\mathbf{k'}$ in Fig. 66 may appear from the preceding argument to be a purely mathematical construction. This cannot be entirely so since, as we have said, equations (45) are the von Laue conditions that determine the quite physical reflection of a wave by the lattice.

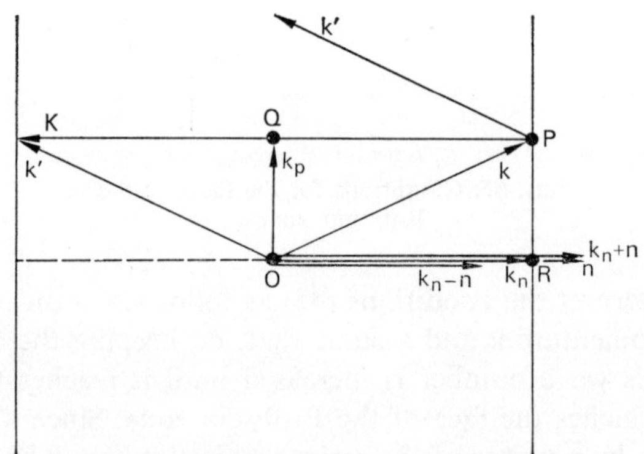

FIG. 67. Reflection of a free wave at the Brillouin zone boundary.

We shall now illustrate the physical meaning of the construction of Fig. 66 by showing that a *free* wave is in fact reflected when its **k** vector reaches the surface of the Brillouin zone. Decompose **k** (Fig. 67) in its components $\mathbf{k}_n$ and $\mathbf{k}_p$ normal and parallel respectively to the Brillouin zone face. We can repeat for $\mathbf{k}_n$ the argument used in § 3.12 [see Fig. 34 (p. 9)] to show that $E(\mathbf{k}_n+\mathbf{n}) = E(\mathbf{k}_n-\mathbf{n})$ whence, for $\mathbf{n} \to 0$ it follows by continuity that $\partial E/\partial \mathbf{k}_n = 0$. On the other hand, the group velocity $\mathbf{v}_g = h^{-1}\,\partial E/\partial \mathbf{k}$ will decompose into its normal and parallel components, $(\mathbf{v}_g)_n$ and $(\mathbf{v}_g)_p$ respectively, the first of which is $(\mathbf{v}_g)_n = h^{-1}\,\partial E/\partial \mathbf{k}_n$ and therefore, from the condition just proved, vanishes, whereas $(\mathbf{v}_g)_p$ is unaltered. This is exactly what happens to the components of the velocity of a ball thrown at an angle against a wall at the moment of impact: if **k** were the momentum of such a ball it would change into **k'** and this must also happen to the free wave.

Warning. The above argument, which is valid for a free (or more precisely an almost free) wave, is *not correct in general,* although it may appear so to the reader.

That there must be a flaw in this argument for the general case can be seen as follows. In $\mathbf{v}_g = h^{-1}\,\partial E/\partial \mathbf{k}$ we ought to compute the partial derivatives at the point P (Fig. 67), that is $\partial E/\partial \mathbf{k}_n$ should be computed *along the line QP.* In order to show from Fig. 67 that $\partial E/\partial \mathbf{k}_n = 0$ along OR we used the centre of inversion at O (compare with Fig. 34), but unless OQ lies on a plane of symmetry (which is by no means always the case, see § 13), there is no reason why $\partial E/\partial \mathbf{k}_n$, computed properly along QP, should vanish. On the other hand, for a free wave, $\partial E/\partial \mathbf{k}_n$ does not depend on $\mathbf{k}_p$ so that if it is zero for $\mathbf{k}_p = 0$ (i.e. along OR), it must always be zero, and hence the proof given above is valid, i.e. the normal component of the group velocity for the vector **k** of Fig. 67 vanishes. (That $\partial E/\partial \mathbf{k}_n$ does not depend on $\mathbf{k}_p$ for a free wave can be easily seen:

$$E = h^2\mathbf{k}^2/2m = h^2\,(\mathbf{k}_p^2+\mathbf{k}_n^2)/2m,$$

so that $\partial E/\partial \mathbf{k}_n = h^2\mathbf{k}_n/m$, independent of $\mathbf{k}_p$.)

Even when one realizes that the argument we gave before is not true in general, it is not easy to see where it fails, since if we decompose the momentum **k** into $\mathbf{k}_n$ and $\mathbf{k}_p$ it appears quite right to define component group velocities in terms of $\mathbf{k}_n$ and $\mathbf{k}_p$, which would then be $\partial E/\partial \mathbf{k}_n$ and $\partial E/\partial \mathbf{k}_p$ computed along OR and OQ respectively. The difficulty is a subtle and important one: what we want to do, really, is to decompose the *velocity*, *not* the momentum. If the velocity and the momentum are along the same direction, as any decent velocities and momenta are, there is no trouble. However, we shall see in § 16 that the group velocity *is parallel to the momentum only for free waves*. This surprising result shows exactly where our argument goes wrong in the general case: we attempted to compute $\mathbf{v}_n$ by decomposing $\mathbf{k}$ (Fig. 68) into $\mathbf{k}_n$ and $\mathbf{k}_p$. But in the same manner that $\mathbf{k}$ is not parallel to $\mathbf{v}$, there is no reason why the group velocity corresponding to $\mathbf{k}_n$ should be along the same axis: i.e. there is no reason why the group velocity corresponding to $\mathbf{k}_n$ should be $\mathbf{v}_n$, that is, normal to the face.

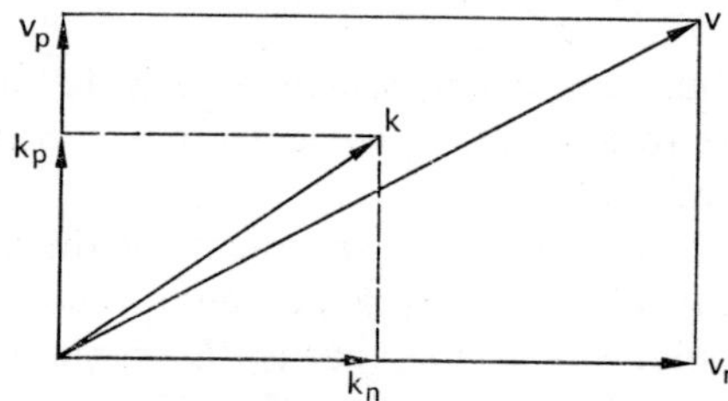

FIG. 68. The velocity and momentum
are not parallel: consequences.

13. Symmetry elements and the faces of the Brillouin zone ($\to$ 17)

The presence of elements of symmetry of the Brillouin zone has certain implications with respect to the vanishing of $\partial E/\partial \mathbf{k}_n$, the derivative of the energy with respect to $\mathbf{k}$ normal to the face.

We have seen in § 8 that $E(\mathbf{k}) = E(-\mathbf{k})$ whether or not the centre O of the Brillouin zone is a centre of inversion of the crystal lattice. It follows from this that $\partial E/\partial \mathbf{k}_n = 0$ whenever $\mathbf{k}$ is normal to a Brillouin zone face from O. This can readily be seen from Fig. 69a, since $E(\mathbf{k}+\mathbf{n}) = E[-(\mathbf{k}+\mathbf{n})]$ from the inversion symmetry mentioned and $E[-(\mathbf{k}+\mathbf{n})] = E(\mathbf{k}-\mathbf{n})$ by periodicity. Therefore $E(\mathbf{k}+\mathbf{n}) = E(\mathbf{k}-\mathbf{n})$ and from continuity $\partial E/\partial \mathbf{k}_n = 0$ for the particular value of $\mathbf{k}$ defined.

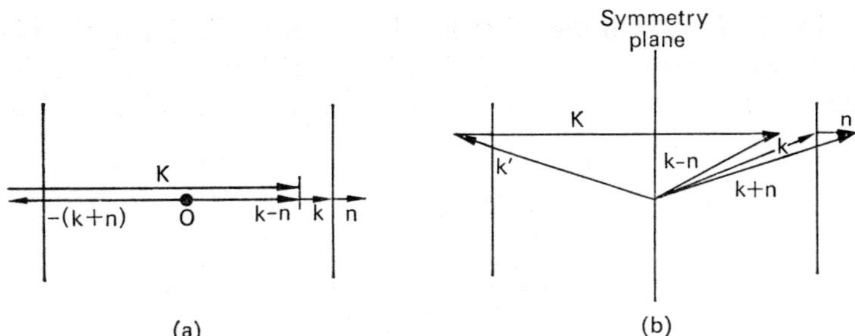

Fig. 69. Properties of $\partial E/\partial \mathbf{k}$.

The reader should verify that the proof given is not valid for a vector $\mathbf{k}$ not normal to the face, such as the one shown in Fig. 69b.

However, $\partial E/\partial k_n$ can under certain conditions vanish for some general values of $\mathbf{k}$ on the Brillouin zone faces, as shown in the Exercises below.

Exercise 1. Plane of symmetry

Show that, if there is a plane of symmetry parallel to a pair of Brillouin zone faces, $\partial E/\partial k_n$ vanishes for each $\mathbf{k}$ that ends on them.

Method. In Fig. 69b, $E(\mathbf{k}+\mathbf{n}) = E(\mathbf{k}')$ by reflection symmetry and $E(\mathbf{k}') = E(\mathbf{k}-\mathbf{n})$ by periodicity, whence $E(\mathbf{k}+\mathbf{n}) = E(\mathbf{k}-\mathbf{n})$ and the result follows.

Exercise 2. Binary axes

Show that if there is a binary axis parallel to a pair of Brillouin zone faces $\partial E/\partial k_n = 0$ for any vector that ends on the line of intersection between these faces and a plane normal to them through the binary axis.

Method. The reader must satisfy himself that, if the central vertical line of Fig. 69b is now a binary axis, the proof given in Exercise 1 is valid for the $\mathbf{k}$ vector shown in the figure but not for any other that ends on the face at points above or below the lines shown in the drawing.

Example. We show in Fig. 70 a reciprocal lattice in two dimensions and the corresponding Brillouin zone, which has no symmetry plane parallel to its faces. On the other hand, the perpendicular to the plane of the drawing through the origin is a binary axis and $\partial E/\partial \mathbf{k}_n$ vanishes for the vectors that end at the points marked in the figure with small dots.

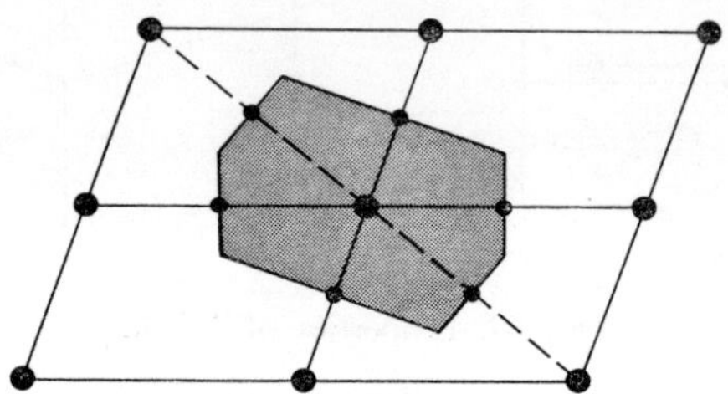

Fig. 70. A Brillouin zone without symmetry planes.
The Brillouin zone is shaded.

14. Energy contours, filling-in of zones, and Fermi surfaces

It should be clear from the treatment in Chapter **2** that the energy of a free electron, in our present notation, is given by $E = h^2\mathbf{k}^2/2\mathbf{m}$. Therefore it depends only on the modulus of the radius vector from the origin in $\mathbf{k}$ space and the corresponding energy surfaces are spherical. As in the one-dimensional case (see § **3**.13) we expect, for nearly free electrons, to follow the free-electron approximation near the centre of the Brillouin zone but to have to depart from it at the edges. As an example, let us consider the first Brillouin zone of a two-dimensional square lattice. This contains two states per atom (§ **11**). If we have a metal with one electron per atom the zone will start to fill in from the lowest energy state at $\mathbf{k} = 0$ outwards, until exactly one half of its volume is filled (see § **11**) as shown in Fig. 71. If we accept that the free-electron approximation is valid up to this point the energy contour corresponding to the highest occupied state is still a circle (or a sphere in the three-dimensional case) the area of which satisfied the condition $\pi r^2 = \frac{1}{2}|\mathbf{a}^*|^2$, from which its radius r is given by $r = |\mathbf{a}^*|(2\pi)^{-\frac{1}{2}}$.

The boundary of the occupied states in $\mathbf{k}$ space must be a surface of constant energy, in fact one for which the energy is E_F, the Fermi energy. Such a surface is called the *Fermi surface* and it is in general of a more or less complicated shape. For free-electrons, of course, it must be a sphere.

Just as in the one-dimensional case, the distortions of the energy contours for nearly free electrons can be most conveniently studied in extended space. We represent in Fig. 72 the first and second Brillouin zones of a square lattice (see Fig. 63) for which $a = 5\,\text{Å}$. Therefore $|\mathbf{a}^*| = 0\cdot2\,\text{Å}^{-1}$ as shown in the figure. The energy contours have been

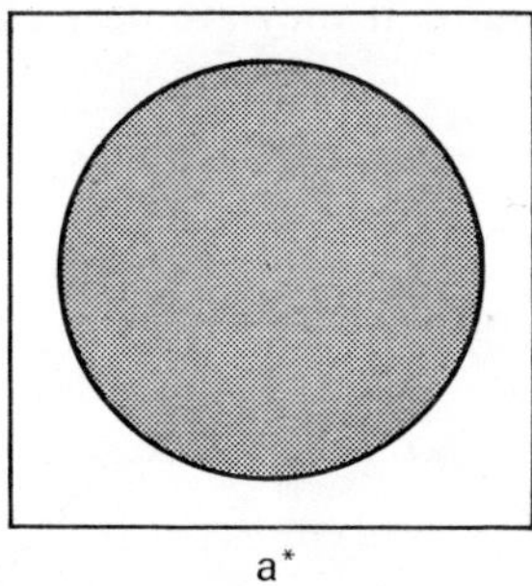

FIG. 71. The Fermi surface for one electron per atom in the two-dimensional square lattice. Notice how large the occupied region of the Brillouin zone looks. This is even more so for the three-dimensional cubic lattice, for which $r = 0\cdot49|\mathbf{a}^*|$, so that the Fermi surface for one electron per atom is almost the inscribed sphere.

obtained from the free-electron circles for which the energy is proportional to the square of their radii. It can be seen that, in rydberg units of energy $E = 11\cdot04\,\mathbf{k}^2$ where $\mathbf{k}$ is measured in Å^{-1} (see Exercise 1 below). In this lattice, the faces of the Brillouin zone are parallel to a plane of symmetry, so that from Exercise 1 in § 13 it follows that for values of $\mathbf{k}$ on the edge of the Brillouin zone the condition $\partial E/\partial \mathbf{k}_n = 0$ must be satisfied, which requires a modification of the hitherto spherical energy contours. (Remember that in the nearly free electron approximation the free-electron results are followed as far as it is permitted by formal conditions such as the one just invoked.) We have guessed in the figure the modifications of the contours required so as to satisfy the vanishing of the derivative although, of course, such changes depend in practice on the potential field of the lattice and would have to be computed. It can be seen that the vanishing of the derivative at the boundary creates discontinuities in the energy. For instance, the contour $E = 0.14$ ryd joins the contour $E = 0.17$ ryd, so that in crossing the surface of the Brillouin zone at the point cut by the $E = 0.14$ ryd contour there is an energy jump of 0.03 ryd. Even those contours that do not cut the Brillouin zone faces are pulled towards them, as shown in the figure, if they are near enough to the boundary.

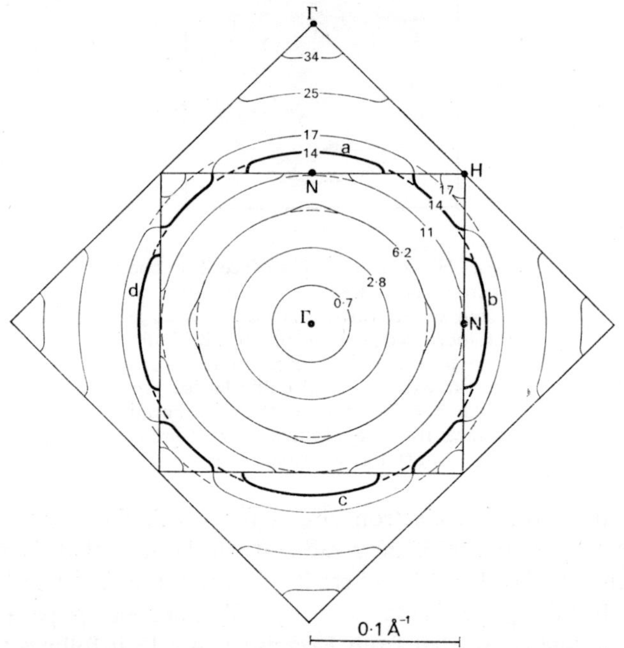

FIG. 72. Energy contours in the first two zones of a square lattice. The energies are given in hundredths of a rydberg.

Suppose now that the metal in question has two electrons per atom. The circle $E = 0.14$ ryd has been drawn so that its area is equal to that of the Brillouin zone and therefore it contains exactly two electrons. On account of the distortions discussed, this circle has to be modified and the thick contour line of the figure ($E = 0\cdot14$ ryd) is such that it still surrounds an area equal to the Brillouin zone, it contains two electrons and is therefore the Fermi surface.

It should be noticed that the first band starts to fill in from the centre (state of lowest energy) outwards and that, although there is an energy gap at the faces, it is more economical in energy to surmount it than to fill in the energy states near the corners of the first zone, which are higher in energy. (This, of course, would no longer be the case if the deviations from the free-electron picture were larger, when the whole of the first Brillouin zone might be filled in before any state of the second is occupied.)

As a result, the second band starts to fill in at N, and when the second zone is folded back on to the first it is clear that it is filled from the edge at N towards the centre at Γ. When this happens, it is often said that there is an *overlap* from the first band on to the second at the centre of the Brillouin zone face. (See § 15.)

It follows from our example that the first zone (first band) is almost full with small pockets of holes at its corners, whereas the second zone (second band) contains small pockets of electrons at the face centres. It should be noticed that, although there are two electrons per atom, the first zone is not full, so that the highest occupied Brillouin zone (the second) is not full either, from which it follows that the material is not an insulator. (See § 15.)

In order to represent the Fermi surface obtained, it is customary to do so band by band and to use the periodically repeated scheme (see § 3.16) as we do in Fig. 73 for the first band. All that we do in this case is to repeat periodically, as shown in Fig. 72, the first Brillouin zone with the contour of the Fermi surface given in it. The result is displayed as on the left of Fig. 74, where one can see that the Fermi surface for the first band consists of pockets of holes at the corners of the Brillouin zone. In order to represent the second band we must fold back the second zone, in which it was given, onto the first one. This is done in the manner shown in Fig. 63 and once this is done the first zone is again repeated periodically. The result is shown on the right of Fig. 74 where, to facilitate identification, the segments of the Brillouin zone which are folded back from those shown in Fig. 72 are labelled with the same letters. It should be clearly understood that both bands in

Fig. 74 are plotted in the first Brillouin zone and that the pieces of the Fermi surface that stick out of the zone must be interpreted in the periodically repeated scheme.

This scheme is extremely advantageous, if not essential, if we want to follow the path of the point in **k** space that represents a changing

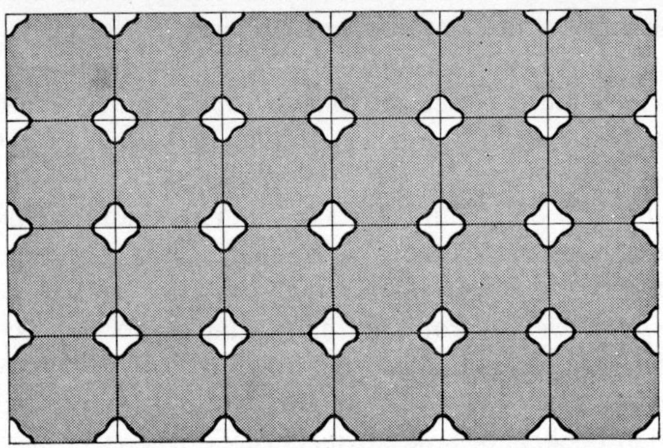

FIG. 73. Periodically repeated scheme for the first band Fermi surface of Fig. 72. The shaded areas contain electrons.

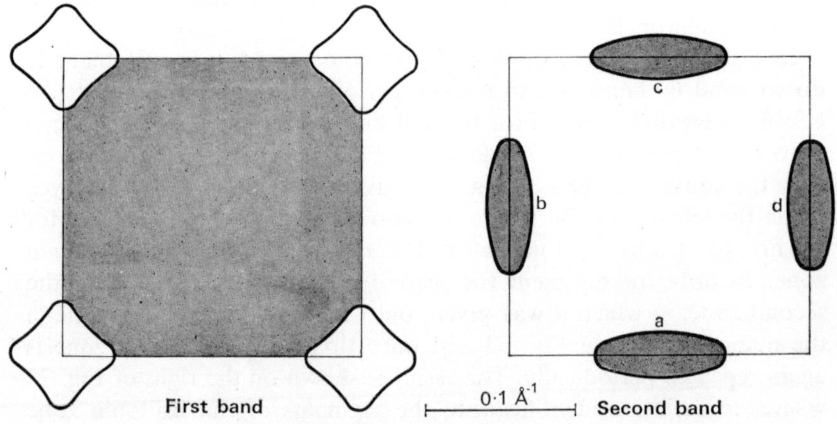

FIG. 74. The first and second band Fermi surfaces of Fig. 72. The shaded areas contain electrons.

electron state. Such a path is called an *orbit* (§ 3.16) and it is particularly interesting to consider orbits along which the energy of the particle is kept constant since, when this constant value is that of the Fermi energy these orbits are experimentally accessible and provide us with very useful information about the shape of the Fermi surface (see § **6.7.2**). Such orbits must, of course, lie on the Fermi surface, and in the two-dimensional case which we are using as an example they will just coincide with the Fermi surface itself. We show in Fig. 75 the orbit that corresponds to the pocket of holes of the first band in Fig. 74,

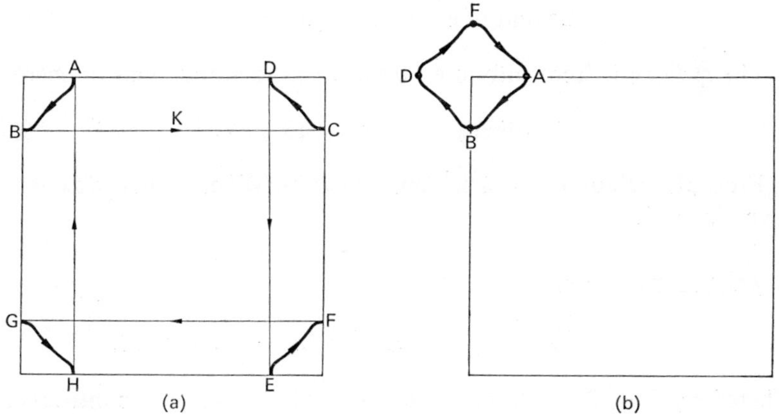

FIG. 75. An orbit in the reduced scheme (a) and in the periodically repeated scheme (b).

and it can be seen that, in the reduced scheme used on the left of the figure (a), the orbit is discontinuous. In fact, if we start at *A* in the figure and move to *B*, the state is reflected from this point to *C*, identical to *B* since it differs from *B* by the **K** vector shown. After *C*, the state changes as follows: $C \to D \to E \to F \to G \to H \to A$. Exactly the same process is represented on the right of the figure (b) in the periodically repeated scheme in a continuous orbit. (It might in fact be easier to look at this orbit first and understand from it how the discontinuous one on the left of the figure arises.)

Exercise 1. Atomic and Rydberg units

The *unit of length* in atomic units is the Bohr radius

$$a_0 = \frac{h^2}{4\pi^2\,\mathbf{m}\mathbf{e}^2} = 0\cdot529\,\text{Å}, \tag{46}$$

where h is Planck's constant and $\mathbf{m}$ and $\mathbf{e}$ the electronic mass and charge respectively.

The *unit of mass* in atomic units is the electron mass $\mathbf{m}$ and the *unit of energy* that of a system of two electrons separated by a_0:

$$\text{atomic unit of energy} = \mathbf{e}^2/a_0 = 27\cdot2 \text{ eV.}$$

The *rydberg* is frequently used instead of the atomic unit of energy:

$$1 \text{ rydberg} = \mathbf{e}^2/2a_0 = 13\cdot6 \text{ eV.} \tag{47}$$

From the relation $E = h^2\mathbf{k}^2/2\mathbf{m}$, find E in rydberg units when $|\mathbf{k}|$ is given in Å^{-1}.

Method. From (46)

$$h^2/2\mathbf{m} = 2\pi^2\mathbf{e}^2 a_0 = (2\pi a_0)^2\mathbf{e}^2/2a_0 = (2\pi a_0)^2 \text{ ryd.}$$

Therefore, $E = (2\pi a_0)^2\mathbf{k}^2$ ryd. When the value of a_0 in Å is introduced here, $|\mathbf{k}|$ must be measured in Å^{-1} and

$$E = 11\cdot04\,\mathbf{k}^2 \text{ ryd} \quad (|\mathbf{k}| \text{ in } \text{Å}^{-1}). \tag{48}*$$

[When circular $\mathbf{k}$ space is used a factor of $1/4\pi^2$ should be included on the right-hand side of (48).]

PROBLEM 1. Verify the energies of the contour lines of Fig. 72 from the free-electron formula (48) with values of $k = 0$, $0\cdot025$, $0\cdot050$, etc.

PROBLEM 2. Find the value in rydbergs of the free-electron energy eigenvalues for the points Γ, H, and N of Fig. 72 in the first band, and of the point Γ in the second band.

PROBLEM 3. Calcium is face centred cubic with lattice constant $a = 5\cdot56\,\text{Å}$. Find its free-electron Fermi energy in rydbergs.

Hints. The unit cell of the face centred cubic lattice is the cube shown in Fig. 78 (p. 174) which contains four atoms. Its reciprocal (which is a cube of edge $a^* = 5\cdot56^{-1}\,\text{Å}^{-1}$) admits two electrons per unit cell (§ 11), i.e. half an electron per atom. The Fermi sphere contains two electrons and therefore must have four times the volume of this cubic reciprocal cell. Deduce from this that the radius k_F of the Fermi sphere is $k_F = \sqrt[3]{(3/\pi)}\,a^*$ and substitute in (48).

PROBLEM 4. Derive from the properties of the Brillouin zone equation (2.39) for the Fermi energy of free electrons.

Hints. (i) Calculate the radius k_F of the (spherical) Fermi surface for a simple cubic lattice with one electron per unit cell and lattice constant a. (ii) From k_F calculate $E_F = h^2 k_F^2/2\mathbf{m}$.

Remarks. Notice that, since we are dealing with free electrons, the crystal lattice is purely fictitious and we are entitled to choose the simplest possible, as we have done. It will be seen that this method is considerably simpler than that used in Chapter 2 as well as being simpler than other methods we shall use later, where the details of the real lattice are used. (Problem 2, § 6.8.3.)

PROBLEM 5. Derive, from the properties of the Brillouin zone, equation (2.46) for the density of states of free electrons.

Hints. The free-electron energy eigenvalues in the range $E, E + \Delta E$ correspond to a spherical shell of radius k and thickness Δk. Call $\nu(\Delta k)$ the number of states in this shell. Then, from equation (2.43), $n(E) = \Omega^{-1}\,\nu(\Delta k)/\Delta E$, where Ω is the volume of the unit cell. $\nu(\Delta k) = 4\pi k^2 \Delta k \,.\, 2\Omega^{*-1}$ (since the volume Ω^* of the first Brillouin zone contains two electrons per atom). Therefore $n(E) = 8\pi k^2 dk/dE$. Replace k^2 and dE/dk by the values that result from $E = h^2 k^2/2\mathbf{m}$ and the result follows.

●●● *Exercise 2: The density of states* (→ **5.3**)

(A knowledge of the properties of the gradient vector is required in this Exercise.)

Prove that the density of states $n(E)$ for electrons of both spins is

$$n(E) = 2 \int_{\Sigma} \frac{d\sigma}{|\mathrm{grad_k}\, E|},$$

where $d\sigma$ is the element of area of the level surface in **k** space Σ for which the energy has the constant value E, the surface integral is taken over the whole of Σ, and $\mathrm{grad_k}\, E$ is the gradient of E in **k** space.

Method. Call Σ' the constant energy surface for which the energy is $E+\Delta E$. As in Problem 5, you must find the volume ω between Σ and Σ' and then the number of states in this shell is $\nu(\Delta k) = \omega\, 2\Omega^{*-1}$. If $\Delta\sigma$ is a small element of area on Σ, $\Delta\omega = \Delta\sigma\Delta k$, where Δk is the perpendicular distance from Σ to Σ'. From the definition of the gradient vector, $|\mathrm{grad_k}\, E| = \Delta E/\Delta k$, whence $\Delta\omega = \Delta E\Delta\sigma/|\mathrm{grad_k}\, E|$. Therefore,

$$\omega = E \int_{\Sigma} d\sigma/|\mathrm{grad_k}\, E|.$$

From this equation, $\nu(\Delta k)$ is obtained at once and since $n(E) = \Omega^{-1}\nu(\Delta k)/\Delta E$, (see Problem 5), the main result follows.

15. Bands, overlaps, conductors, and insulators

We found in one dimension that a metal with two electrons per atom would have to be an insulator or at most a semi-conductor, if the energy gap happens to be small. Of course, the physical meaning of this result must be limited, since a one-dimensional model of a metal is not a very realistic one. In fact, we have just seen in the two-dimensional example of § 14 that the rule in question breaks down, because states in the second zone begin to fill in *before* the whole of the first zone is full. This means that there are energy states in the second band below states of the first band, which, in one dimension, would require a—forbidden—band crossing. In other words, although

there are energy discontinuities at the edge of the Brillouin zone *there
is no forbidden energy gap.*

How this result can come about should be much clearer from Fig. 76,
where the bands that correspond to the energy contours of Fig. 72 are
shown. (In practice, bands such as those of Fig. 76 are computed
directly by methods described in Chapter 6. The opposite procedure
is used here because we obtained Fig. 72 by guesswork.) In Fig. 76
the energy bands are plotted along three directions in the Brillouin
zone, which are identified in the inset by means of a customary notation
in band theory whereby characteristic **k** vectors are denoted with
conventional letters, the same letter denoting any of the vectors in a
star. The reader should check that the energy values given in Fig. 76
correspond to those in Fig. 72, although it should be appreciated that,

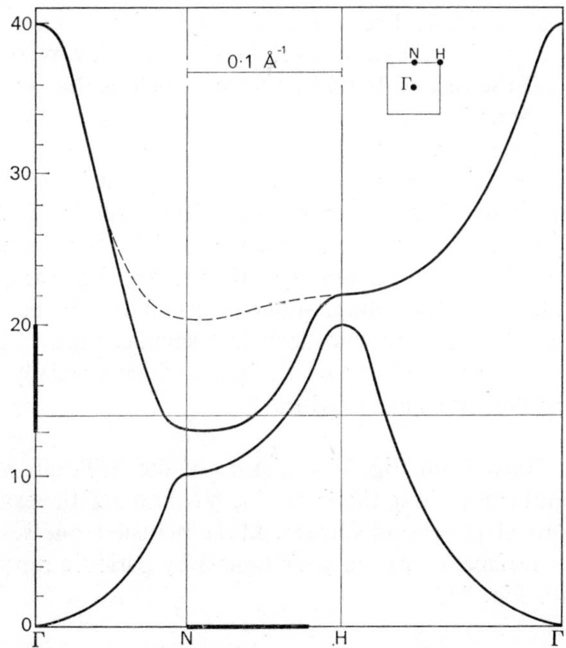

FIG. 76. Energy bands for the two-dimensional square lattice
of Fig. 72. The horizontal line denotes the Fermi energy.
The vertical scale gives the energy in hundredths of a rydberg.

in order to draw the bands, judicious guesses had to be made for the eigenvalues for N and H in the first and second bands and for Γ in the second band. This was done by very rough interpolation and comparison with free-electron eigenvalues, to ensure that the bands of Fig. 76 are plausible.

We notice at once two things when we look at Fig. 76. (i) The first band is empty above the Fermi energy at and around H, i.e. there are holes around H in the first band. The second band is full below the Fermi energy at and around N, i.e. there are pockets of electrons around N in the second band. (ii) For each value of E in the range of energies displayed there is an energy state either in the first or in the second band or in *both*. In fact, there is a part of the edge of the Brillouin zone along NH, marked with a thick line, where the same eigenvalue belongs to two different bands: we say that there is an *energy overlap* on this edge, which starts at N. The range of energies for which there is this overlap is shown with a thick line on the E axis. If this overlap did not exist, that is, if the second band had a form such as that shown in the figure with a dotted line, an *energy gap* would exist and the material would be an insulator.

The basic condition for conduction that we gave in § 3.18, namely that there must be *electron states occupied in a partly filled Brillouin zone*, is still valid: the argument that we used in one dimension can easily be extended to three. On the other hand, the example of this section should show that simple-minded arguments based purely on the number of electrons are worthless since small changes in the energy bands can create or eliminate energy gaps which crucially affect the way in which the bands are filled in.

PROBLEM. Draw from Fig. 76 a picture of the Brillouin zone in the same scale and mark along the lines ΓN, NH and $H\Gamma$ the exact lengths of the sections of the Fermi surface. Make plausible guesses as to the shape of the surface, complete your figures by periodic repetition and compare with Fig. 74.

Remark. It is essentially in this way that the shapes and sizes of Fermi surfaces are derived from band systems such as that of Fig. 76, but obtained from direct computation. See an example in § 6.8.3.

16. Velocity† (12 ← , → **end**)

We have seen in § 1.16 that the electron velocity must be described by means of the group velocity $\mathbf{v}_g = h^{-1}\, dE/dk$. This expression was correct in one dimension: in three we must expect the velocity to be described by three components of the form $(\mathbf{v}_g)_i = (1/h)\, \partial E/\partial k_i$, $(i = x, y, z,)$. We use here reciprocal lattice suffices since the components of $\mathbf{k}$ are always given in these axes. Correspondingly, the components of the velocity just given must be understood as components in reciprocal or $\mathbf{k}$ space. It follows at once from the definition of the gradient vector that $\mathbf{v}_g$ is the gradient of E in reciprocal space:

$$\mathbf{v}_g = \frac{1}{h}\, \mathrm{grad}_\mathbf{k}\, E, \tag{49}$$

where the suffix $\mathbf{k}$ in the grad denotes that it should be taken in reciprocal rather than direct space.

A well known property of the gradient vector is that it is perpendicular to the level surfaces of the scalar quantity for which it is computed. Hence $\mathbf{v}_g$ is perpendicular to the level surfaces of the energy: since the Fermi surface is a level surface it follows that the velocity of the electrons at the Fermi surface is perpendicular to it.

We see in Fig. 77a that, for spherical Fermi surfaces, $\mathbf{v}_g$ is parallel to $\mathbf{k}$, which means that for free electrons the velocity and the momentum are effectively the same thing except for a constant factor, a result which was proved for one-dimensional cases in (1.66). On the other hand, in Fig. 77b, we see that $\mathbf{v}_g$ is no longer parallel to $\mathbf{k}$ for non-spherical Fermi surfaces, and the distinction between the velocity and the momentum is significant, a result which was discussed in § 12.

The above considerations provide a direct physical meaning for the Fermi surface: it is the surface to which the electron velocity is always normal. This makes the Fermi surface a concept of primary importance in the study of transport phenomena of electrons in lattices.

† A knowledge of the properties of the gradient vector is required in this section.

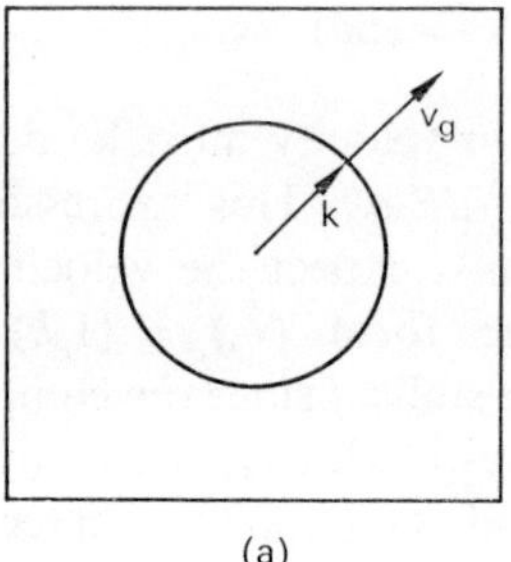
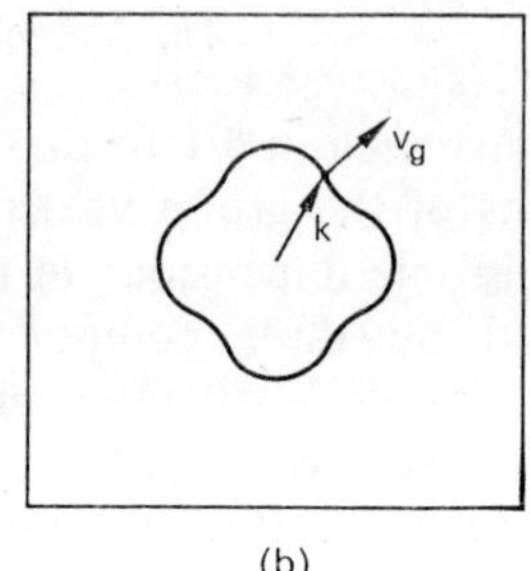

(a)   (b)

FIG. 77. The velocity and momentum are not parallel.

17. The Brillouin zone for cubic lattices

So far, our examples of Brillouin zones have been one- or two-dimensional. As a three-dimensional example we shall obtain the Brillouin zone for the face centred cubic lattice, which is represented in Fig. 78, where $\mathbf{u}, \mathbf{v}, \mathbf{w}$ are the primitive vectors of the lattice.

We must first find the reciprocal lattice, for which we observe in the figure that $\mathbf{u}^*$ must be perpendicular to the plane determined by the vectors $\mathbf{v}$ and $\mathbf{w}$. This plane, which is the large triangle in the figure, is a diagonal plane of the cube, so that $\mathbf{u}^*$ must be along the diagonal axis perpendicular to it, as shown in the figure (notice in it that we do not attempt to represent the moduli of the reciprocal vectors in the proper scale). In the same manner $\mathbf{v}^*$ and $\mathbf{w}^*$ are found along the other diagonals and it can be seen at once that the lattice generated by the reciprocal lattice vectors is a body centred cubic one. We conclude that *the reciprocal lattice of a face centred cubic lattice is a body centred cubic lattice.*

Clearly, the centred unit cell of the body centred lattice will be the Brillouin zone sought. In order to obtain it we must draw the perpendicular bisector planes to the lines that join the point at the origin in $\mathbf{k}$ space to its first neighbours. These are of two types, labelled 1 and 2 in Fig. 79. Point 2 is at the centre of the neighbouring cell, so that the bisector plane lies on the top face of the cube shown. Also, there is fourfold symmetry around the line $\varGamma 2$, so that the Brillouin zone face perpendicular to it must be a polygon of a number of sides equal to

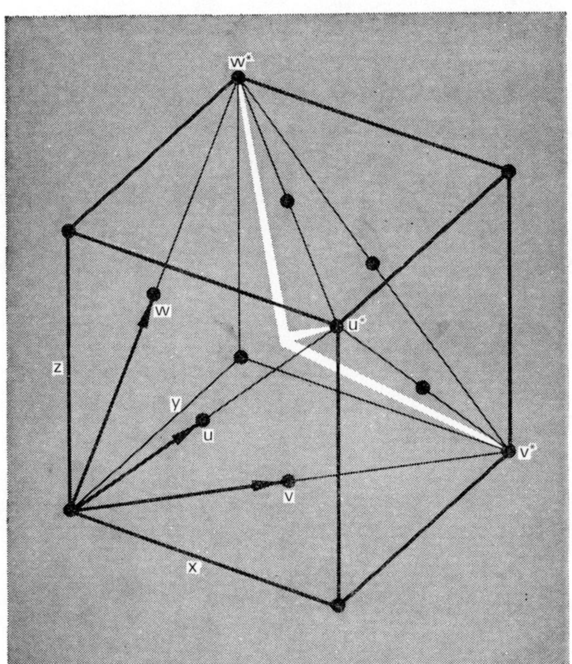

FIG. 78. Face centred cubic lattice and its
reciprocal vectors. The lattice constant *a* is
the length of the side of the cube.

four or a multiple of it. The bisector of the line $\Gamma 1$ must be inside the
cube on a plane parallel to a diagonal plane of the cube: on account
of the threefold symmetry around it the corresponding face must be a
polygon of a number of sides equal to three or a multiple. When the
various possibilities are examined it can be seen that the faces of the
first type are squares and those of the second regular hexagons, as
shown in the figure. It should be noticed, of course, that the axis
perpendicular to an hexagonal face is not sixfold but threefold. In
fact, it can be seen in the figure that, as discussed in § 7, the reciprocal
lattice as well as the Brillouin zone have exactly the same cubic symmetry
as the direct lattice.

1. Properties of the Faces of the F.C.C. Brillouin zone (→ **6.**8)

The faces of the Brillouin zone differ in more than in shape. The square faces are parallel to symmetry planes and therefore $\partial E/\partial \mathbf{k}_n$ vanishes for **k** vectors ending on them. (See Exercise 1, § 13.) The hexagonal faces are parallel to diagonal planes of the cube, which are not symmetry planes. However, there are binary axes parallel to these faces, one of which is shown in Fig. 79. This is the binary axis of the

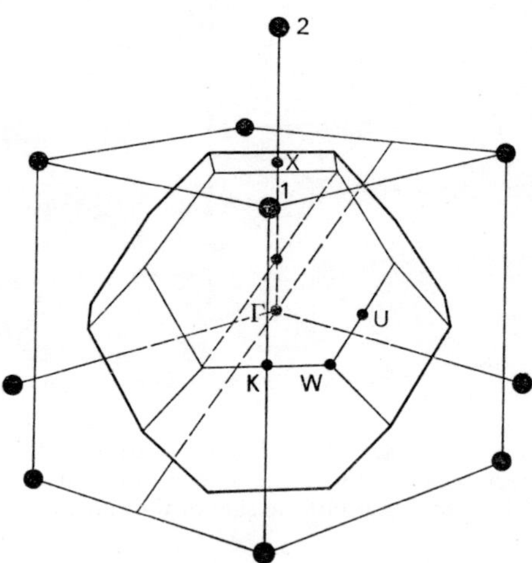

FIG. 79. The first Brillouin zone for the face centred cubic lattice. The edge of the cube is equal to $2a^{-1}$. (See Fig. 80.)

cube that joins the midpoints of two opposite edges and which is also, of course, a binary axis of the Brillouin zone. It follows from the result of Exercise 2, § 13, that $\partial E/\partial \mathbf{k}_n = 0$ for all **k** vectors that end on the axes of the hexagonal face that join opposite vertices. One of these axes is shown in the picture. This result will be of importance when discussing the Fermi surface of copper in § 6.8.3.

Exercise 1

Show, in the manner of the text, that the reciprocal lattice of a b.c.c. lattice is a f.c.c. one. (In fact, it should be clear from general principles that if the reciprocal of a lattice A is B, that of B must be A.)

● *Exercise 2*

(This exercise requires a knowledge of vector products and mixed triple products.)

Show by vector methods that the reciprocal of a f.c.c. lattice is a b.c.c. one and that if a is the lattice constant of the f.c.c. lattice that of its b.c.c. reciprocal is $2a^{-1}$.

Method. In what follows $[mnp]$ denotes a vector with components m, n, p along the x, y, z axes of the f.c.c. lattice of Fig. 78. Attention must be paid to the units.

In units of a, it follows from Fig. 78 that the primitive vectors are

$$\mathbf{u} = [\tfrac{1}{2} \ 0 \ \tfrac{1}{2}], \quad \mathbf{v} = [\tfrac{1}{2} \ \tfrac{1}{2} \ 0], \quad \mathbf{w} = [0 \ \tfrac{1}{2} \ \tfrac{1}{2}]. \tag{50}$$

The volume Ω of the primitive cell spanned by these vectors is

$$\Omega = \mathbf{u} \cdot \mathbf{v} \times \mathbf{w} = \begin{vmatrix} \tfrac{1}{2} & 0 & \tfrac{1}{2} \\ \tfrac{1}{2} & \tfrac{1}{2} & 0 \\ 0 & \tfrac{1}{2} & \tfrac{1}{2} \end{vmatrix} = \tfrac{1}{4}$$

or, with units, $\Omega = \tfrac{1}{4}a^3$.

From the formulae (21), $\mathbf{u}^* = \mathbf{v} \times \mathbf{w}/\mathbf{u} \cdot \mathbf{v} \times \mathbf{w}$ and cyclic permutation, verify that

$$\mathbf{u}^* = (1 \ \bar{1} \ 1), \quad \mathbf{v}^* = (1 \ 1 \ \bar{1}), \quad \mathbf{w}^* = (\bar{1} \ 1 \ 1), \tag{51}$$

in units of a^{-1}. (Observe that, in arbitrary units, these vectors agree with those displayed in Fig. 78, where no attempt was made to get right the length of the reciprocal vectors.)

Verify that $\mathbf{u}^* \cdot \mathbf{v}^* \times \mathbf{w}^* = 4$ so that the volume Ω^* of the primitive cell spanned by these vectors is $\Omega^* = 4a^{-3}$. This is, of course, also the volume of the first Brillouin zone shown in Fig. 79. (Notice that $\Omega^* = \Omega^{-1}$ in agreement with Exercise 4, § 4.)

Plot the vectors (51) as in Fig. 80 and notice that they span a b.c.c. lattice with lattice constant $2a^{-1}$.

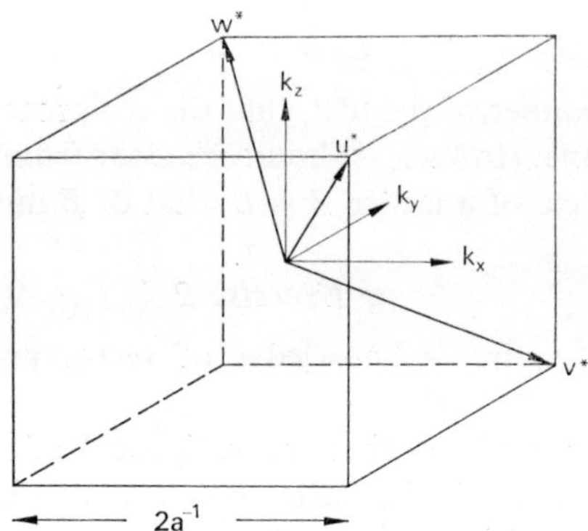

FIG. 80. Primitive vectors of the reciprocal
lattice of the f.c.c. lattice.

● *Exercise 3*

Verify the values found for Ω and Ω^* from first principles, by considering the *unit cells*, rather than the primitive cells (i.e. by considering the lattices as simple cubic lattices with a basis inside the unit cell).

Method. (i) We had $\Omega = \frac{1}{4}a^3$. a^3 is the volume of the unit cell of Fig. 78, which has four atoms per cell. The primitive cell spanned by **u, v, w** has only one atom per cell, so that its volume should be $\frac{1}{4}$ of that of the unit cell, as found.

(ii) If we consider the f.c.c. lattice as a simple cubic lattice with four atoms per unit cell, then $\Omega = a^3$ and $\Omega^* = \Omega^{-1} = a^{-3}$ is the volume of the corresponding cell in the reciprocal lattice. However, since there are four atoms in the direct lattice, there are four times as many electrons as in the corresponding simple cubic lattice and four of these cells should be filled, with a total volume of $4a^{-3}$ which is the value that we found in Exercise 2.

(iii) The reciprocal lattice that we found is b.c.c. with lattice constant $2a^{-1}$. The volume of its unit cell (shown in Fig. 80) is therefore $8a^{-3}$. This unit cell has two lattice points per cell (a property of the b.c.c. lattice) so that the primitive cell (with one lattice point per cell) must have half this volume, i.e. $4a^{-3}$, again in agreement with the value found.

Exercise 4

It is customary to designate some important **k** vectors with conventional letters, such as, for the f.c.c. lattice, Γ, X, L, K, W in Fig. 79. Verify that, in units of a^{-1}, the components of these vectors are

$$\Gamma = [0\ 0\ 0],\ \ X = [0\ 0\ 1],\ \ L = [\tfrac{1}{2}\ \tfrac{1}{2}\ \tfrac{1}{2}],\ \ K = [\tfrac{3}{4}\ \tfrac{3}{4}\ 0],\ \ W = [\tfrac{1}{2}\ 1\ 0]\quad (52)^*$$

Method. The first three values are immediate from Fig. 79 when it is remembered that the edge of the cube shown in it is $2a^{-1}$ (see Fig. 80).

It should be noticed that the symbols given for the **k** vectors in (52) are also used for **k** vectors obtained from them by symmetry, that is on the same star (see § 7) although their components may be different. Thus all central points of the square faces are called X. In Fig. 81 we give the horizontal section of the Brillouin zone of Fig. 79. The point X at the top is the one on the square face on the right-hand side of Fig. 79. The point X underneath belongs to the bottom face of another Brillouin zone that shares with the Brillouin zone of Fig. 79 the hexagonal face that contains L. We know that $\Gamma X = a^{-1}$ and that $XW = WX$, hence $XW = \tfrac{1}{2}a^{-1}$ and the coordinates of W follow at once: likewise those of K.

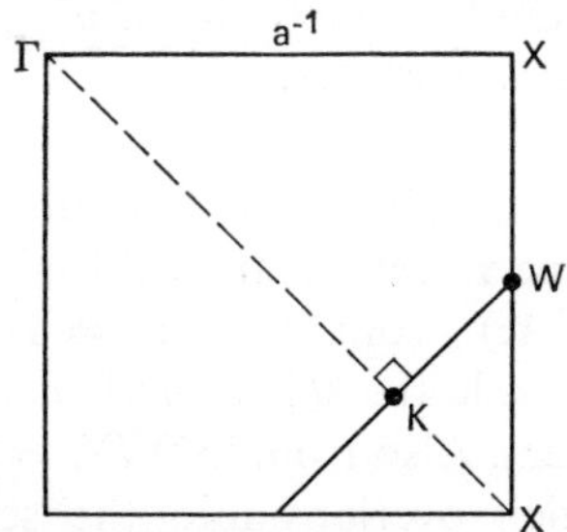

Fig. 81. Horizontal section of the
Brillouin zone of Fig. 79.

PROBLEM 1. Show that, for the f.c.c. lattice, the radius of the sphere that corresponds to an electron concentration of one electron per atom (Brillouin zone half full) is about 0·9 of the radius ΓL.

Hint. From (52) $\Gamma L = \tfrac{1}{2}\sqrt{(3)}a^{-1}$. The radius sought is defined by $\tfrac{4}{3}\pi t^3 = \tfrac{1}{2}\Omega^* = 2a^{-3}$.

G

Notice that even for a half-full Brillouin zone the free-electron Fermi surface gets extremely near the centre of the hexagonal face, a result which will be important in dealing with the Fermi surface of copper (§ 6.8).

PROBLEM 2. Verify that the free-electron Fermi sphere that touches the face-centred cubic Brillouin zone at L corresponds to an electron concentration of $1\cdot36$. [You should notice from (52) that L is the point of the Brillouin zone which is nearest to the origin.]

Hint. The ratio of the volume of the sphere of radius ΓL to the volume of the first Brillouin zone is $\frac{4}{3}\pi(\Gamma L)^3/4a^{-3}$. Since one Brillouin zone contains two electrons, this ratio has to be multiplied by 2 in order to obtain the number of electrons contained in the inscribed sphere.

PROBLEM 3. Show that the free-electron Fermi surface for a f.c.c. metal with two electrons per atom has pockets of holes in the first band going from K to W and pockets of electrons in the second band around L. Draw a picture to scale (units of a^{-1}) of the $\Gamma X W K$ section and $\Gamma X K L$ section of these surfaces.

Hint. The radius of the sphere corresponding to two electrons per atom (sphere of the same volume as the Brillouin zone) is $0\cdot99$. (Notice that it almost touches X.) From (52) $\Gamma W = 1\cdot1$, $\Gamma K = 1\cdot06$, so that these points are left outside the sphere and there is a sort of torus of holes going through them. Also from (52) $\Gamma L = 0\cdot866$ so that L is well inside the sphere, which overlaps into the second zone around L forming a calotte outside it.

PROBLEM 4. Calcium is f.c.c. with lattice constant $a = 5\cdot56\,\text{Å}$. Calculate in the free-electron approximation the lowest eigenvalues for calcium at Γ, K, X, L, W in rydbergs. From the value of the Fermi energy derived for this metal in § 14, Problem 3, deduce which of these points will be occupied and which will be empty in the first band. Compare this result with those of Problem 3.

18. Hexagonal close-packed lattice. Jones zones

We show in Fig. 82a the disposition of the layers in the hexagonal closed packed (h.c.p.) lattice, and in (b) its well known unit cell. The h.c.p. is a lattice with basis and it should be clear that the unit cell contains two lattice points (there are eight at the corners of the unit cell but each point is shared by eight cells, which leaves just one corner point to which the internal one must be added). We know (§ 3.21) that the reciprocal lattice is obtained without consideration of the

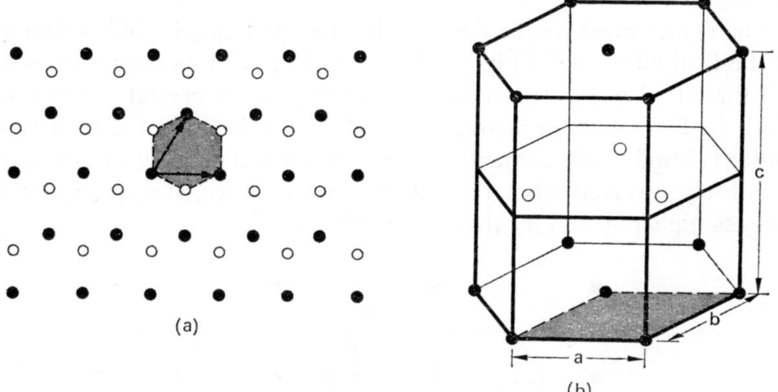

FIG. 82. The hexagonal close-packed lattice. The solid circles correspond to the first and third layers and the open one to the intermediate layer. The unit cell is the prism with shaded base in (b). For later purposes, notice that this unit cell is equivalent to another, the base of which is the shaded area in (a), which is equal to the shaded area in (b).

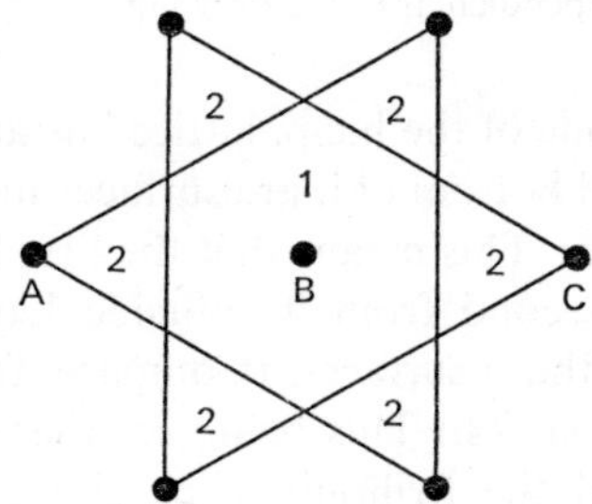

FIG. 83. The first two Brillouin zones for a hexagonal close-packed layer.

contents of the unit cell, i.e. of the intermediate layer in Fig. 82a. We have seen in Fig. 56 (p. 136) that the reciprocal lattice of a h.c.p. layer is itself a close-packed layer, and we show in Fig. 83 the construction of the first two Brillouin zones for it. In the three-dimensional case the successive layers of the reciprocal lattice overlap one another exactly. Hence, the hexagon of Fig. 83 will be the horizontal sections of the first Brillouin zone, and vertical sections of the Brillouin zones can be obtained from Fig. 84. From Figs. 83 and 84 it can be seen that the first Brillouin zone must be an hexagonal prism as shown in Fig. 85, and that the basic features of the second Brillouin zone agree with those given in the figure.

The reader must not be deceived by the apparent six-fold symmetry of the Brillouin zone of Fig. 85. Its vertical axis can have no more than the three-fold symmetry of the vertical axis of the crystal lattice (Fig. 82b). In fact, it was remarked in p. 147 that because the reciprocal lattice is blind to the contents of the unit cell and because its symmetry is entirely received from that of the direct lattice, it can appear to possess higher symmetry than really obtains.

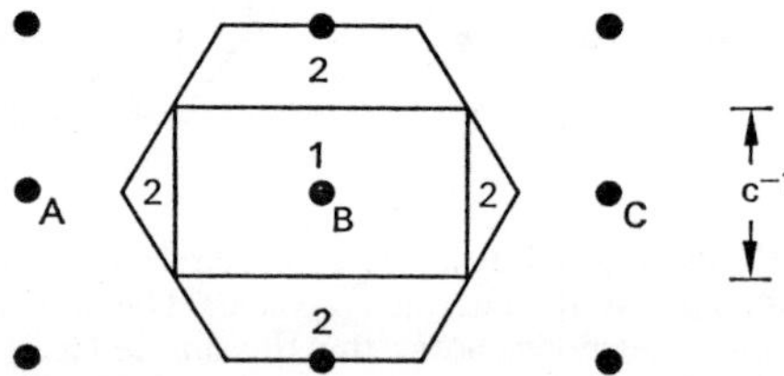

FIG. 84. The first and second Brillouin zones for the h.c.p. lattice. A section is shown on a plane perpendicular to the drawing of Fig. 83 through *A, B, C*.

The first Brillouin zone of the h.c.p. lattice has an interesting feature, namely that the top and bottom of it are surfaces upon which the energy cannot be discontinuous. This means that the first band from the inside of the zone and the second from its outside have exactly the same energy eigenvalues on those surfaces. (Compare, for example, with the opposite situation in Fig. 72.) This being so, these surfaces are not of much significance, and the Brillouin zone should more properly be continued across them. This feature will be discussed sketchily in this section and rigorously in § 19. An alternative explanation will also be given in § 22.5.

The three atoms shown in the middle of the cell of Fig. 82 form a h.c.p. layer which is staggered with respect to those above and below it. If this middle layer were shifted so as to be superimposed with the other two, the period along the vertical direction would become $\frac{1}{2}c$ instead of c. Correspondingly, the vertical axis of the first Brillouin zone (see Fig. 84) would be duplicated in length from c^{-1} to $2c^{-1}$. It turns out that, from the point of view of the $E(\mathbf{k})$ surfaces, the effect of the intermediate layer is the same as if it were superimposed on the others. This means that the top and bottom of the Brillouin zone should more properly be taken as those of the second Brillouin zone (see Fig. 84) which are, in fact, separated by $2c^{-1}$.

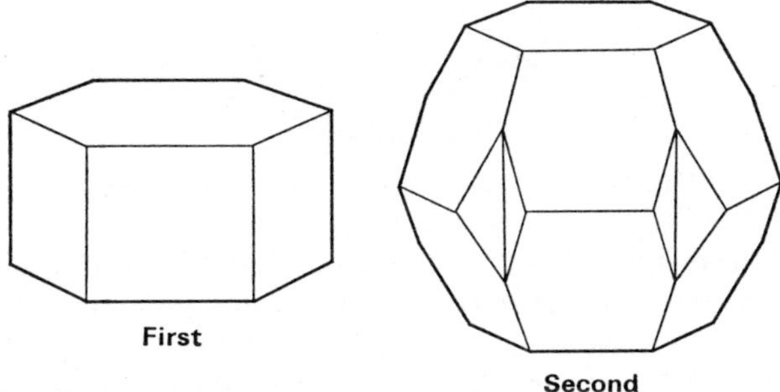

FIG. 85. The first and second Brillouin zone of the h.c.p. lattice.

The meaning of this result is the following: the top and bottom of the first Brillouin zone in Fig. 85 are no longer surfaces of discontinuity, although the lateral faces retain this property. Since the surfaces of discontinuity are physically significant, in particular in making predictions as to possible changes of the free-electron bands, it is sometimes useful to construct a cell in $\mathbf{k}$ space such that the whole of its surface corresponds to discontinuities of the energy. In order to do this for the h.c.p. lattice we can keep the sides of the first Brillouin zone of Fig. 85, but we must replace its top and bottom by the top and bottom parts of the second Brillouin zone. We construct in this way the zone shown in Fig. 86, in which the horizontal faces are separated by $2c^{-1}$.

Such a zone, entirely bounded by surfaces of discontinuity, is called a *Jones zone*. The reader should beware of the fact that in the older literature (until the middle fifties or so) no distinction in name was made between the Jones and the Brillouin zones.

The Brillouin zone always contains two electrons per unit cell of the crystal lattice (which corresponds to one electron per atom in the h.c.p. lattice, for which there are two atoms in the unit cell of the crystal lattice). A Jones zone, on the other hand, can contain a variable number of electrons, since its volume is no longer fixed and equal to that of the

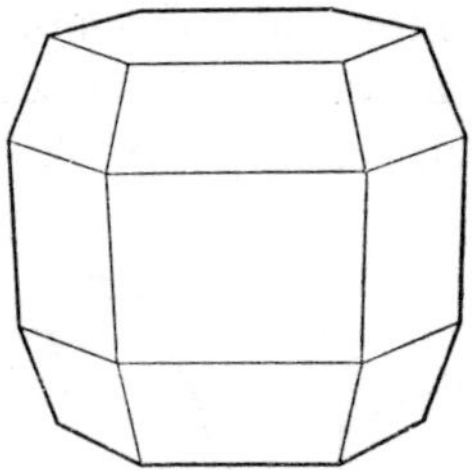

FIG. 86. The Jones zone for the h.c.p. lattice.

unit cell of the reciprocal lattice. In order to find this number it is enough to obtain the ratio of the volume of the Jones zone and that of the Brillouin zone. A geometrical study of the polyhedron shown in Fig. 86 shows that it contains the whole of the first Brillouin zone and very approximately three-quarters of the second, so that it will contain about 1·75 states per atom.

●● 19. Sticking together of bands in the h.c.p. lattice (→ end)

We shall now discuss in more detail why the top and bottom faces of the h.c.p. lattice are not surfaces of energy discontinuities.

In order to do this we must first notice the existence of an important symmetry element in the h.c.p. lattice, which is called a *screw rotation*. In Fig. 87 we show the unit cell of the h.c.p. lattice that corresponds to the shaded hexagon of Fig. 82a. (The reader should verify that this cell contains two atoms, the six of the bases being shared each by six cells and the three of the central plane being shared by three.) The axis of this figure is a *binary screw axis,* the corresponding covering operation being defined as follows: rotate by π around the vertical axis *and* translate by $\frac{1}{2}c$ in a direction parallel to this axis. The reader should notice that the existence of this unusual symmetry operation depends on the fact that we have a lattice with basis, i.e. with more than one atom per unit cell: if the middle layer in Fig. 87 did not exist

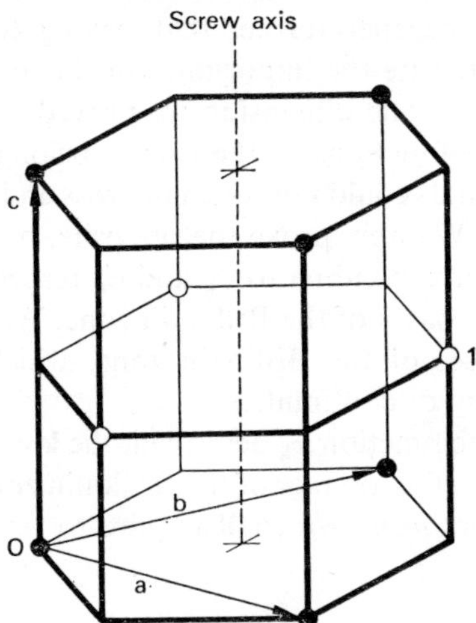

FIG. 87. A unit cell of the h.c.p. lattice.
Screw rotation.

a screw rotation would not be possible. We shall represent this screw rotation with the symbol **C**. A very important property of it is that

$$\mathbf{C}^2 = \mathbf{R}, \tag{53}$$

where **R** is a translation by c. It follows in fact from Fig. 87 that when **C** is repeated twice the bottom layer is brought exactly into coincidence with the top one.

We must consider the effect of **C** on a vector **k**. So far we have seen that symmetry operations such as rotations drag the reciprocal lattice together with the direct lattice and thereby effect symmetry transformations of the **k** vectors. Translations, on the other hand, cannot affect **k** vectors since all they do is to displace the Brillouin zone as a whole. Since **C** is the product of a binary rotation (rotation by $2\pi/2$) **C$_2$** with a translation by $\frac{1}{2}c$, the effect of **C** on a **k** vector is exactly the same as that of **C$_2$**.

We shall now discuss the properties of Bloch functions $\varphi_\mathbf{k}$, where **k** is any value on one of the bases of the Brillouin zone. We shall prove that for each such value of **k** there belong *two* wave functions $\varphi_\mathbf{k}$ and $\varphi_\mathbf{k}'$, say, which are degenerate, i.e. that belong to the same energy eigenvalue. To appreciate the importance of this result remember that for the linear chain in one dimension we proved that there cannot be two degenerate eigenfunctions of the energy belonging to the same **k**, which meant that bands could not touch or cross and led to the existence of the energy gaps. We now prove the opposite, namely that there are two bands, those corresponding to $\varphi_\mathbf{k}$ and $\varphi_\mathbf{k}'$ respectively which touch *on every point* of the bases of the Brillouin zone. We say that they *stick together* on the bases of the Brillouin zone, which therefore cannot correspond to energy discontinuities.

Let us consider the function $\varphi_\mathbf{k}$ described, the **k** value of which is represented in Fig. 88 on the top face of the Brillouin zone: $\mathbf{k} = [m, n, \frac{1}{2}c^{-1}]$ for any real numbers m, n. We shall require to know the effect of **R**,

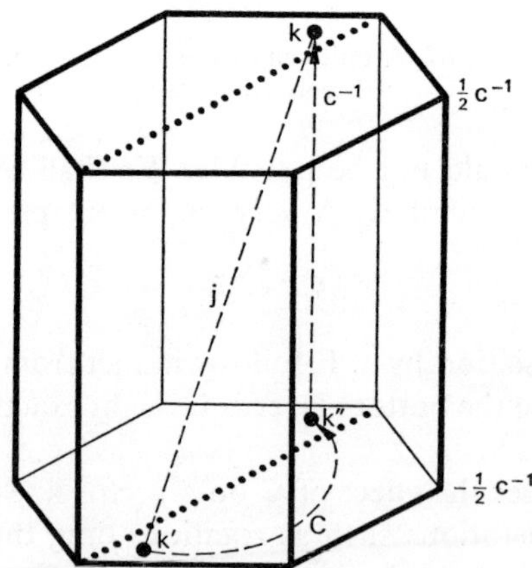

Fig. 88. Effect of **Cj** on **k**.

the translation by c that appears in (53). Since $\mathbf{R} = [0\ 0\ c]$, $\mathbf{k} \cdot \mathbf{R} = \frac{1}{2}$, so that

$$\mathbf{R}\varphi_{\mathbf{k}} = \exp(2\pi i \mathbf{k} \cdot \mathbf{R})\varphi_{\mathbf{k}} = \exp(\pi i)\varphi_{\mathbf{k}} = -\varphi_{\mathbf{k}}. \tag{54}$$

In order to prove the result in question we must prove two things. Firstly, that there is a function $\varphi'_{\mathbf{k}}$ that corresponds to the same energy as $\varphi_{\mathbf{k}}$. This is most simply done by proving that there is a function which is derived from $\varphi_{\mathbf{k}}$ by acting on it with symmetry operations and which corresponds to the same value of $\mathbf{k}$. We call this function $\varphi'_{\mathbf{k}}$. Secondly, we must prove that $\varphi'_{\mathbf{k}}$ is independent of $\varphi_{\mathbf{k}}$. In fact, in order to establish the existence of degeneracy it is required that $\varphi_{\mathbf{k}}$ and $\varphi'_{\mathbf{k}}$ belong to the same energy *and* be independent (see § **1.11**):

$$\varphi'_{\mathbf{k}} \neq \alpha\varphi_{\mathbf{k}} \tag{55}$$

(α a constant).

We fulfil the first part of this programme in Fig. 88, where $\mathbf{j}$ is the complex conjugation operation (§ 8) that inverts $\mathbf{k}$ through the origin. Clearly,

$$\mathbf{k} \overset{j}{\to} \mathbf{k}' \overset{c}{\to} \mathbf{k}'' \equiv \mathbf{k}, \tag{56}$$

where $\mathbf{k}'' \equiv \mathbf{k}$ since they differ by a vector of the reciprocal lattice $[0\ 0\ c^{-1}]$.

The compound operation $\mathbf{Cj}$ (remember from § **3.2** that this means that $\mathbf{j}$ is effected first and $\mathbf{C}$ second) has therefore the following effect on $\varphi_{\mathbf{k}}$:

$$\mathbf{Cj}\varphi_{\mathbf{k}} = \text{function that belongs to } \mathbf{k}$$

$$= \varphi'_{\mathbf{k}}, \tag{57}$$

say.

We must now fulfil the second part of our programme. In fact, suppose that $\varphi'_{\mathbf{k}}$ is dependent on $\varphi_{\mathbf{k}}$, thus, negating (55),

$$\varphi'_{\mathbf{k}} = \alpha\varphi_{\mathbf{k}}. \tag{58}$$

From (57),

$$\mathbf{Cj}\varphi_{\mathbf{k}} = \alpha\varphi_{\mathbf{k}}. \tag{59}$$

Operate with $\mathbf{Cj}$ on both sides of this equation. Since $\mathbf{j}$ is *not* linear (see § 8) $\mathbf{Cj}$ is not linear either:

G*

$$\mathbf{Cj}(\alpha\varphi_k) = \mathbf{Cj}\alpha\mathbf{Cj}\varphi_k = \mathbf{j}\alpha\mathbf{Cj}\varphi_k = \alpha^*\mathbf{Cj}\varphi_k. \tag{60}$$

(In the penultimate step here we write $\mathbf{Cj}\alpha = \mathbf{j}\alpha$ since α is a constant, i.e. a number, complex in general, that cannot be affected by an ordinary symmetry operation. $\mathbf{j}$, on the other hand, transforms it into its complex conjugate α^*.) Therefore, from (59) and (60),

$$\mathbf{CjCj}\varphi_k = \alpha^*\mathbf{Cj}\varphi_k$$
$$= \alpha^*\alpha\varphi_k, \tag{61}$$

on using (59) again.

We know from Exercise 2, § 8, that $\mathbf{j}$ commutes with any symmetry operation. Also, $\mathbf{jj} = \mathbf{E}$, the identity, since taking the complex conjugate twice in succession is to return to the original value. Therefore in (61)

$$\mathbf{CjCj}\varphi_k = \mathbf{Cjj}\mathbf{C}\varphi_k = \mathbf{C}^2\varphi_k = \mathbf{R}\varphi_k, \tag{62}$$

where in the last step we use (53). On replacing this result in (61),

$$\mathbf{R}\varphi_k = \alpha^*\alpha\varphi_k \tag{63}$$
$$= -\varphi_k, \tag{64}$$

on account of (54). From (63) and (64),

$$\alpha^*\alpha = -1, \tag{65}$$

which is impossible. (Write $\alpha = u + iv$, then $\alpha^*\alpha = u^2 + v^2$ which must always be positive.) It follows that there does not exist an α that satisfies (58), whence (55) is true and φ_k' and φ_k are independent.

PROBLEM. Reproduce the same proof for a simple hexagonal lattice (one for which the intermediate layer in Fig. 87 is missing). Notice that $\mathbf{C}$ is no longer a symmetry operation. Carry out the proof with $\mathbf{C}_2$ in lieu of $\mathbf{C}$. You should find that now φ_k' and φ_k are dependent and that there is no degeneracy. That is, the existence of the screw axis $\mathbf{C}$ is responsible for the sticking together of the bands in the h.c.p. lattice.

● 20. Fourier series in three dimensions

In the rest of this chapter we shall extend to three dimensions the one-dimensional work done in §§ 3.22–3.24.

1. THE ORTHONORMAL FUNCTIONS

We saw there that a periodic function

$$V(x) = V(x+a), \quad 0 \leqslant x \leqslant L, \tag{66}$$

can be expanded in terms of the functions

$$\varphi_m(x) = a^{-\frac{1}{2}} \exp(2\pi i m x a^{-1}), \tag{67}$$

which are orthonormal in the periodic interval $0 < x < a$. We are now interested in triply periodic functions

$$V(xyz) = V(x+a, y+b, z+c), \quad x, y, z \text{ in volume } v, \tag{68}$$

and in order to expand them we shall verify that the functions

$$\varphi_{mnp}(xyz) = \Omega^{-\frac{1}{2}} \exp[2\pi i(mxa^{-1} + nxb^{-1} + pxc^{-1})], \tag{69}$$

which are an extension of those given by (67), are orthonormal in the unit cell of volume Ω spanned by the vectors $\mathbf{a}, \mathbf{b}, \mathbf{c}$. If we call $\mathbf{K}_\nu$ the vector of the reciprocal lattice

$$\mathbf{K}_\nu \equiv \mathbf{K}_{mnp} = m\mathbf{a}^* + n\mathbf{b}^* + p\mathbf{c}^*,$$

and $\mathbf{r}$ the vector defined by (1), (69) can be written as follows:

$$\varphi_\nu(\mathbf{r}) = \Omega^{-\frac{1}{2}} \exp(2\pi i \mathbf{K}_\nu \cdot \mathbf{r}). \tag{70}$$

We must prove that

$$I_{\nu\nu'} = \int_\Omega \varphi_\nu^* \varphi_{\nu'} \, d\omega = \delta_{\nu\nu'}, \tag{71}$$

where $d\omega$ is the volume element corresponding to the axes chosen in the unit cell. Clearly,

$$I_{\nu\nu'} = \Omega^{-1} \int_{\Omega} \exp[2\pi i(\mathbf{K}_{\nu'} - \mathbf{K}_{\nu}) \cdot \mathbf{r}] \, d\omega. \tag{72}$$

We shall carry out the proof in two parts:

(i) $\nu = \nu'$. $\quad I_{\nu\nu} = \Omega^{-1} \int_{\Omega} d\omega = \Omega^{-1}\Omega = 1.$ \hfill (73)

(ii) $\nu \neq \nu'$. The components of $\mathbf{K}_{\nu'} - \mathbf{K}_{\nu}$ are the integers $M = m' - m$, $N = n' - n$, $P = p' - p$. Therefore

$$I_{\nu\nu'} = \Omega^{-1} \int_{\Omega} \exp[2\pi i(Mxa^{-1} + Nyb^{-1} + Pzc^{-1})] d\omega. \tag{74}$$

This will factorize into three integrals, over x, y, z respectively. Consider the first, I_x, say,

$$I_x = \int_0^a \exp(2\pi i \, Mxa^{-1}) \, dx$$

$$= \frac{a}{2\pi i M} \exp(2\pi i \, Mxa^{-1})\Big|_0^a. \tag{75}$$

The exponential here is equal to unity for $x = 0$ and for $x = a$ (since M is integral). Therefore $I_x = 0$ and similar results are valid for I_y and I_z whence $I_{\nu\nu'} = 0$ for $\nu \neq \nu'$. This result combined with (73) verifies (71).

Exercise. Free electron eigenfunctions

Reproduce the derivation of the free-electron eigenfunctions (2.27) and (2.28) given in § 2.6 for a cubic sample of metal, for the more general case of an orthogonal parallelepiped of edges $L_x = N_x a$, $L_y = N_y b$, $L_z = N_z c$ (N_x, N_y, N_z integers), and volume v, to show that

$$\psi(xyz) = v^{-\frac{1}{2}} \exp\left[2\pi i\left(\frac{\kappa_x}{N_x}\frac{x}{a} + \frac{\kappa_y}{N_y}\frac{y}{b} + \frac{\kappa_z}{N_z}\frac{z}{c}\right)\right], \tag{76}$$

which, from (1) and (7), can be rewritten as

$$\psi_{\mathbf{k}}(\mathbf{r}) = v^{-\frac{1}{2}} \exp(2\pi i \mathbf{k} . \mathbf{r}). \tag{77}$$

These free-electron eigenfunctions are normalized over the whole volume v of the crystal, whereas the functions (70) are normalized over the unit cell. Except for this minor difference, notice that the orthonormal functions (70) are the free-electron eigenfunctions (77) for the particular case when their $\mathbf{k}$ vectors coincide with vectors of the reciprocal lattice.

2. Expansion of the periodic functions

$V(xyz)$ of (68), now rewritten $V(\mathbf{r})$, will be expressed as a series in terms of the functions $\varphi_\nu(\mathbf{r})$:

$$V(\mathbf{r}) = \sum_\nu V_\nu \varphi_\nu(\mathbf{r}). \tag{78}$$

The coefficients V_ν are obtained as follows, from (78):

$$\int_\Omega \varphi^*_{\nu'}(\mathbf{r}) V(\mathbf{r}) \, d\omega = \sum_\nu V_\nu \int_\Omega \varphi^*_{\nu'}(\mathbf{r}) \varphi_\nu(\mathbf{r}) \, d\omega = \sum_\nu V_\nu \delta_{\nu\nu'} = V_{\nu'}. \tag{79}$$

In full notation, (78) and (79) are as follows:

$$V(\mathbf{r}) = \Omega^{-\frac{1}{2}} \sum_\nu V_\nu \exp(2\pi i \mathbf{K}_\nu . \mathbf{r}), \tag{80}$$

$$V_\nu = \Omega^{-\frac{1}{2}} \int_\Omega \exp(-2\pi i \mathbf{K}_\nu . \mathbf{r}) V(\mathbf{r}) \, d\omega. \tag{81}$$

Equation (80) is the *Fourier series* for $V(\mathbf{r})$ and the coefficients V_ν are the *Fourier coefficients*.

● 21. Lattice with basis. Structure factor

Consider now $V(\mathbf{r})$ to be the potential function in a crystal lattice. If there is only one atom per unit cell $V(\mathbf{r})$ is expanded straightaway as in (80). However, we want to consider the more general case, when there are several atoms in the unit cell at positions designated with vectors $\mathbf{u}_j$ $(j = 1, 2, \dots)$, the components of which, like those of $\mathbf{r}$ are given as dimensionless quantities in terms of a, b, and c. Call

$$\mathbf{r}_j = \mathbf{r} - \mathbf{u}_j. \tag{82}$$

The potential at the point $\mathbf{r}$ due to the atom at $\mathbf{u}_j$ will be $V(\mathbf{r}_j)$. Hence, the total potential at $\mathbf{r}$ will be

$$V(\mathbf{r}) = \sum_j V(\mathbf{r}_j). \tag{83}$$

Clearly, $V(\mathbf{r}_j)$ is a periodic function in the lattice which can be expanded in a Fourier series like (80):

$$V(\mathbf{r}_j) = \Omega^{-\frac{1}{2}} \sum_v V_{v,j} \exp\left(2\pi i \mathbf{K}_v \cdot \mathbf{r}_j\right), \tag{84}$$

where $V_{v,j}$ is the vth Fourier coefficient of the potential of atom j. On introducing (82) into (84),

$$V(\mathbf{r}_j) = \Omega^{-\frac{1}{2}} \sum_v V_{v,j} \exp\left(-2\pi i \mathbf{K}_v \cdot \mathbf{u}_j\right) \exp\left(2\pi i \mathbf{K}_v \cdot \mathbf{r}\right), \tag{85}$$

and when this value is introduced in (83) we obtain

$$V(\mathbf{r}) = \Omega^{-\frac{1}{2}} \sum_v \mathscr{V}_v \exp\left(2\pi i \mathbf{K}_v \cdot \mathbf{r}\right), \tag{86}$$

where

$$\mathscr{V}_v = \sum_j V_{v,j} \exp\left(-2\pi i \mathbf{K}_v \cdot \mathbf{u}_j\right) \tag{87}$$

is called the *structure factor*. If there is only one atom per unit cell, which can always be taken at the origin ($\mathbf{u}_1 = 0$), $\mathscr{V}_v = V_v$. In fact, on comparing (86) with (80) it follows that all that is required in order to expand the potential of a lattice with basis is to replace the Fourier coefficient of the potential V_v by the structure factor $\mathscr{V}_v$.

The structure factor has a very important bearing on the energy gaps at the faces of the Brillouin zone and we shall see in § 22.5 how they are computed.

PROBLEM. If $\mathscr{V}_v$ is the structure factor that pertains to $\mathbf{K}_v$ in accordance to (87) call $\mathscr{V}_{-v}$ the one that corresponds to $-\mathbf{K}_v$. Show, on using (87) and (81) that, for real $V(\mathbf{r})$, $\mathscr{V}_{-v} = \mathscr{V}_v^*$.

● 22. The nearly free-electron approximation

We shall extend and generalize here the one-dimensional treatment of the *nearly free-electron approximation* (NFE) given in § 3.23.

1. THE BASIC EXPANSION

We assume that the effective crystal hamiltonian $^e\mathbf{H}(\mathbf{r})$ (see § 2.1) differs very little from the free-electron hamiltonian $\mathbf{H}(\mathbf{r})$:

$$^e\mathbf{H}(\mathbf{r}) = \mathbf{H}(\mathbf{r}) + V(\mathbf{r}), \tag{88}$$

where the crystal potential $V(\mathbf{r})$ is assumed small. Therefore, the free-electron eigenfunctions $\psi_{\mathbf{k}}(\mathbf{r})$ from (77) are good approximations to the crystal eigenfunctions $\Psi_{\mathbf{k}}(\mathbf{r})$. We represent these by the variational expansion

$$\Psi_{\mathbf{k}}(\mathbf{r}) = \sum_i c_{\mathbf{k}_i}\psi_{\mathbf{k}_i}(\mathbf{r}) \tag{89}$$

with

$$\psi_{\mathbf{k}_i}(\mathbf{r}) = v^{-\frac{1}{2}}\exp(2\pi i\mathbf{k}_i . \mathbf{r}). \tag{90}$$

Of course $\psi_{\mathbf{k}}(\mathbf{r})$ must be one of the functions that appears in (89) [it is likely, in fact, to provide the largest contribution to it, as the function most similar to $\Psi_{\mathbf{k}}(\mathbf{r})$ with which it coincides in the limit $V(\mathbf{r}) \to 0$].

In what follows we shall obtain a restriction on the functions $\psi_{\mathbf{k}_i}(\mathbf{r})$ appearing in (89). The secular determinant (**1.78**) can be written as

$$\det|H_{\mathbf{k}_i\mathbf{k}_j} - \varepsilon S_{\mathbf{k}_i\mathbf{k}_j}| = 0. \tag{91}$$

Since $S_{\mathbf{k}_i\mathbf{k}_j} = \delta_{\mathbf{k}_i\mathbf{k}_j}$ [see (71) as applied to the functions (77)], we have

$$\det|H_{\mathbf{k}_i\mathbf{k}_j} - \varepsilon\,\delta_{\mathbf{k}_i\mathbf{k}_j}| = 0. \tag{92}$$

Suppose now that we choose the *m* functions $\psi_{\mathbf{k}_i}(\mathbf{r})$ in such a way that they fall out into two sets one for $i = 1, \ldots, n$ another for $i = n+1, \ldots, m$, such that, whenever we mix two functions of different sets, their matrix elements vanish. (92) will therefore take the form

$$\begin{vmatrix}
H_{\mathbf{k}_1\mathbf{k}_1}-\varepsilon & H_{\mathbf{k}_1\mathbf{k}_2} & \cdots & H_{\mathbf{k}_1\mathbf{k}_n} & 0 & \cdots & 0 \\
H_{\mathbf{k}_2\mathbf{k}_1} & H_{\mathbf{k}_2\mathbf{k}_2}-\varepsilon & \cdots & H_{\mathbf{k}_2\mathbf{k}_n} & 0 & \cdots & 0 \\
\vdots & \vdots & & \vdots & \vdots & \cdots & \vdots \\
H_{\mathbf{k}_n\mathbf{k}_1} & H_{\mathbf{k}_n\mathbf{k}_2} & \cdots & H_{\mathbf{k}_n\mathbf{k}_n}-\varepsilon & 0 & \cdots & 0 \\
0 & 0 & \cdots & 0 & H_{\mathbf{k}_{n+1}\mathbf{k}_{n+1}}-\varepsilon & \cdots & H_{\mathbf{k}_{n+1}\mathbf{k}_m} \\
0 & 0 & \cdots & 0 & H_{\mathbf{k}_{n+2}\mathbf{k}_{n+1}} & \cdots & H_{\mathbf{k}_{n+2}\mathbf{k}_m} \\
\vdots & \vdots & \cdots & \vdots & \vdots & \cdots & \vdots \\
0 & 0 & \cdots & 0 & H_{\mathbf{k}_m\mathbf{k}_{n+1}} & \cdots & H_{\mathbf{k}_m\mathbf{k}_m}-\varepsilon
\end{vmatrix}=0. \tag{93}$$

From a simple property of determinants, the above determinantal equation factorizes in two, each of the equations entailing the vanishing of each of the two determinants strung along the diagonal of (93). It then follows quite simply that expansion (89) falls out into two entirely separate parts one for $i = 1, \ldots, n$ and the other for $j = n+1, \ldots, m$ and that it is a sheer waste of time to keep them together. If $\psi_{\mathbf{k}}(\mathbf{r})$ is one of the functions in the first lot, this must provide a good approximation to $\Psi_{\mathbf{k}}(\mathbf{r})$ and the functions of the second lot contribute nothing and should not be considered.

2. Condition for the Matrix Elements

From the above discussion it follows that it is important to find when a matrix element $H_{\mathbf{k}_i\mathbf{k}_j}$ vanishes.

$$H_{\mathbf{k}_i\mathbf{k}_j} = \int_v \psi^*_{\mathbf{k}_i}(\mathbf{H}+V)\psi_{\mathbf{k}_j}\,d\omega, \tag{94}$$

where v is the volume of the crystal. $\psi_{\mathbf{k}_j}$ is an eigenfunction of $\mathbf{H}$:

$$\mathbf{H}\psi_{\mathbf{k}_j} = E_{\mathbf{k}_j}\psi_{\mathbf{k}_j}.$$

Therefore

$$H_{\mathbf{k}_i\mathbf{k}_j} = E_{\mathbf{k}_j}\delta_{\mathbf{k}_i\mathbf{k}_j} + \int_v \psi^*_{\mathbf{k}_i}V\psi_{\mathbf{k}_j}\,d\omega. \tag{95}$$

We expand V in Fourier series (86)

$$V = \Omega^{-\frac{1}{2}} \sum_{\nu} \mathscr{V}_{\nu} \exp(2\pi i \mathbf{K}_{\nu} \cdot \mathbf{r}) \tag{96}$$

and introduce (96) in the integral in (95):

$$H_{\mathbf{k}_i \mathbf{k}_j} = E_{\mathbf{k}_j} \delta_{\mathbf{k}_i \mathbf{k}_j} + \Omega^{-\frac{1}{2}} v^{-1} \sum_{\nu} \mathscr{V}_{\nu} \int_v \exp[2\pi i (\mathbf{K}_{\nu} - \mathbf{k}_i + \mathbf{k}_j) \cdot \mathbf{r}] \, d\omega. \tag{97}$$

As in (72), the integral is zero if the coefficient of $2\pi i$ does not vanish,[†] i.e. if $\mathbf{K}_{\nu} \neq \mathbf{k}_i - \mathbf{k}_j$; otherwise it is equal to v. Therefore

$$H_{\mathbf{k}_i \mathbf{k}_j} = E_{\mathbf{k}_j} \delta_{\mathbf{k}_i \mathbf{k}_j} + \Omega^{-\frac{1}{2}} \sum_{\nu} \mathscr{V}_{\nu} \delta_{\mathbf{k}_i - \mathbf{k}_j, \mathbf{K}_{\nu}}. \tag{98}$$

Consider the case $\mathbf{k}_i \neq \mathbf{k}_j$. Since their difference is unique we either have one vector of the reciprocal lattice, $\mathbf{K}_{\mu}$ say, such that $\mathbf{k}_i - \mathbf{k}_j = \mathbf{K}_{\mu}$ or we have none. In the second case no term survives in the summation of (98):

$$H_{\mathbf{k}_i \mathbf{k}_j} = 0 \quad \text{for} \quad \mathbf{k}_i \neq \mathbf{k}_j \quad \text{and} \quad \mathbf{k}_i - \mathbf{k}_j \neq \mathbf{K}_{\mu} \quad (\text{some } \mu). \tag{99}$$

In the first case, only one term survives in the summation in (98),

$$H_{\mathbf{k}_i \mathbf{k}_j} = \Omega^{-\frac{1}{2}} \mathscr{V}_{\mu} \quad \text{for} \quad \mathbf{k}_i \neq \mathbf{k}_j \quad \text{and} \quad \mathbf{k}_i - \mathbf{k}_j = \mathbf{K}_{\mu}. \tag{100}$$

If $\mathbf{k}_i = \mathbf{k}_j$, $\mathbf{k}_i - \mathbf{k}_j = \mathbf{K}_0$ (the null reciprocal lattice vector), and (98) gives

$$H_{\mathbf{k}_j \mathbf{k}_j} = E_{\mathbf{k}_j} + \Omega^{-\frac{1}{2}} \mathscr{V}_0. \tag{101}$$

3. Form of the expansion

From the discussion so far it follows that a restriction must be established on the values of $\mathbf{k}_i$ that are permitted to appear in the expansion (89) of $\Psi_{\mathbf{k}}(\mathbf{r})$. Since $\psi_{\mathbf{k}}(\mathbf{r})$ must always appear in this expansion, the values of $\mathbf{k}_i$ in (89) must be such that $\mathbf{k}_i - \mathbf{k} = \mathbf{K}_{\nu}$. We therefore write

$$\Psi_{\mathbf{k}}(\mathbf{r}) = \sum_{\nu} c_{\mathbf{K}_{\nu}} \psi_{\mathbf{k} + \mathbf{K}_{\nu}}(\mathbf{r}). \tag{102}$$

---

† The fact that the integral in (97) is over the crystal volume v and that in (72) over that of the unit cell Ω is no worry since v is a multiple of Ω.

In fact all the values of **k** that appear in (102) differ amongst each other by some vector of the reciprocal lattice and lead to non-vanishing matrix elements. On the other hand, a value of **k** such as **k** + **k**′ (**k**′ not a reciprocal lattice vector) will be such that its matrix elements with all the functions in (102) vanish and should therefore not be included.

4. The energy gaps

If enough terms are included in (102) the method that we have developed provides a fairly accurate value of the energy eigenvalues of the successive bands (see, however, § 6.3). In what follows, however, we want to develop the more interesting, semi-quantitative aspects of the method, for which it is enough to restrict ourselves to a very simple expansion, which contains only two values of **k**, **k** itself, and $\mathbf{k} + \mathbf{K}_\mu$. From the matrix elements given in (100) and (101) it follows that the secular determinant is

$$\begin{vmatrix} E_{\mathbf{k}} + \Omega^{-\frac{1}{2}}\mathscr{V}_0 - \varepsilon & \Omega^{-\frac{1}{2}}\mathscr{V}_\mu^* \\ \Omega^{-\frac{1}{2}}\mathscr{V}_\mu & E_{\mathbf{k}+\mathbf{K}_\mu} + \Omega^{-\frac{1}{2}}\mathscr{V}_0 - \varepsilon \end{vmatrix} = 0. \tag{103}$$

[Notice that the element 1, 2 is $H_{\mathbf{k},\,\mathbf{k}+\mathbf{K}_\mu}$ so that, in (100),

$$\mathbf{k}_i - \mathbf{k}_j = \mathbf{k} - (\mathbf{k} + \mathbf{K}_\mu) = -\mathbf{K}_\mu,$$

whence $H_{\mathbf{k},\,\mathbf{k}+\mathbf{K}_\mu} = \Omega^{-\frac{1}{2}}\mathscr{V}_{-\mu} = \Omega^{-\frac{1}{2}}\mathscr{V}_\mu^*$ from the problem in § 21.]

The two roots of (103) are readily found. From our point of view, we shall be interested in their difference, which we shall call $\Delta_{\mathbf{k},\,\mathbf{k}+\mathbf{K}_\mu}$ in order to bear in mind the two values of **k** that are used in the expansion of the wave function.

$$\Delta_{\mathbf{k},\,\mathbf{k}+\mathbf{K}_\mu} = [(E_{\mathbf{k}} + E_{\mathbf{k}+\mathbf{K}_\mu})^2 - 4E_{\mathbf{k}}E_{\mathbf{k}+\mathbf{K}_\mu} + 4\Omega^{-1}|\mathscr{V}_\mu|^2]^{\frac{1}{2}}$$

$$= [(E_{\mathbf{k}} - E_{\mathbf{k}+\mathbf{K}_\mu})^2 + 4\Omega^{-1}|\mathscr{V}_\mu|^2]^{\frac{1}{2}}. \tag{104}$$

If $E_{\mathbf{k}} - E_{\mathbf{k}+\mathbf{K}_\mu}$ is large, an approximate value of the square root can be taken:

$$\Delta_{\mathbf{k},\,\mathbf{k}+\mathbf{K}_\mu} = E_\mathbf{k} - E_{\mathbf{k}+\mathbf{K}_\mu} + \frac{2\Omega^{-1}|\mathscr{V}_\mu|^2}{E_\mathbf{k} - E_{\mathbf{k}+\mathbf{K}_\mu}}, \tag{105}$$

which is approximately equal to $E_\mathbf{k} - E_{\mathbf{k}+\mathbf{K}_\mu}$. Since this is the difference between the energies corresponding to the original functions $\psi_\mathbf{k}$ and $\psi_{\mathbf{k}+\mathbf{K}_\mu}$ it follows that the correction obtained is very small and that although some gain in accuracy may be achieved by the expansion, no new qualitative feature is introduced.

Consider now the opposite case when

$$E_\mathbf{k} - E_{\mathbf{k}+\mathbf{K}_\mu} = 0. \tag{106}$$

Since these free-energy eigenvalues depend only on the moduli of the $\mathbf{k}$ vectors, (106) entails

$$|\mathbf{k}| = |\mathbf{k}+\mathbf{K}_\mu|. \tag{107}$$

If we call $\mathbf{k}' = \mathbf{k}+\mathbf{K}_\mu$, we also have from (107) $|\mathbf{k}'| = |\mathbf{k}|$ and these are the conditions (45) that characterize a $\mathbf{k}$ vector that ends on the surface of a Brillouin zone. The value of Δ that we obtain for this case from (104) gives therefore the energy gap on the surface of the Brillouin zone. This is

$$(\Delta_{\mathbf{k},\,\mathbf{k}+\mathbf{K}_\mu})_{\mathbf{k}\ \text{on Brillouin zone}} = 2\Omega^{-\frac{1}{2}}|\mathscr{V}_\mu|. \tag{108}$$

(Compare with Problem 3, § 3.24.) For a simple lattice $\mathscr{V}_\mu \equiv V_\mu$ and each energy gap is therefore a direct measure of the corresponding Fourier coefficient of the potential. For lattices with bases (108) shows the crucial importance of the structure factor in determining energy gaps, which we shall illustrate in the next section.

5. The Structure Factor and Energy Gaps

We shall compute $\mathscr{V}_\mu$ for the h.c.p. lattice. The two atoms of the unit cell are labelled 0 and 1 in Fig. 87 (the choice of these atoms is of course largely arbitrary). It can be readily verified that

$$u_0 = [0\ 0\ 0], \quad u_1 = [\tfrac{2}{3}\ \tfrac{2}{3}\ \tfrac{1}{2}]. \tag{109}$$

In order to compute the gap at the top face of the Brillouin zone we have to take $\mathbf{K}_\mu = [0\ 0\ 1]$, since this is the $\mathbf{K}$ vector that transfers a vector $\mathbf{k}$ that ends on this face into another vector of the same length (see Fig. 84 and remember that the vector $\mathbf{K} = [0\ 0\ 1]$ corresponds to a length c^{-1} along the vertical direction of the figure).

Since the two atoms of the unit cell are identical, we can take $V_{\mu,\,0} = V_{\mu,\,1} \equiv V_\mu$ in (87), which with the values chosen of u_0, u_1, and $\mathbf{K}$ gives

$$\mathscr{V}_{[0\ 0\ 1]} = V_{[0\ 0\ 1]}\,[1+\exp(-2\pi i\tfrac{1}{2})] = V_{[0\ 0\ 1]}\,(1-1) = 0. \quad (110)$$

This shows from (108) that there is no energy gap on the top and bottom faces of the Brillouin zone of the h.c.p. lattice, a result which we proved otherwise in § 19.

Exercise

Prove that in the f.c.c. lattice

$$\mathscr{V}_{[mnp]} = 4V_{[mnp]} \quad \text{for} \quad m,n,p \text{ all of the same parity,} \quad (111)$$

$$= 0 \quad \text{otherwise.} \quad (112)$$

Method. In Fig. 78 call 0, 1, 2, 3, the atoms at the origin and at the tips of $\mathbf{u}$, $\mathbf{v}$, $\mathbf{w}$ respectively. Verify that

$$u_0 = [0\ 0\ 0],\ u_1 = [\tfrac{1}{2}\ 0\ \tfrac{1}{2}],\ u_2 = [\tfrac{1}{2}\ \tfrac{1}{2}\ 0],\ u_3 = [0\ \tfrac{1}{2}\ \tfrac{1}{2}].$$

From (87)

$$\mathscr{V}_{[mnp]} = V_{[mnp]}\{1+\exp[-\pi i(m+p)]$$

$$+\exp[-\pi i(m+n)]+\exp[-\pi i(n+p)]\}. \quad (113)$$

Call A, B, C the first, second and third exponentials. When m, n, p are all even or all odd, $A = B = C = 1$, whence

$$\mathscr{V}_{[mnp]} = 4V_{[mnp]}.$$

When m is even and n and p are odd, $A = -1$, $B = -1$, $C = 1$, and $\mathscr{V} = 0$, and this result holds for any combination of one even index with two odd ones. Likewise $\mathscr{V} = 0$ for m odd and n, p even or for any combination of one odd and two even indices. This proves the result.

Remark. The **K** vector that separates the top and bottom square faces of the f.c.c. Brillouin zone is [0 0 2] (see legend to Fig. 79). Likewise, from this figure, two hexagonal faces are separated by [1 1 1]. From (111) the corresponding structure factors do not vanish and therefore energy gaps exist on those faces.

PROBLEM. Prove that in the b.c.c. lattice

$$\mathscr{V}_{[mnp]} = 2V_{[mnp]} \quad \text{for} \quad m+n+p = \text{even}, \tag{114}$$

$$= 0 \quad \text{for} \quad m+n+p = \text{odd}. \tag{115}$$

Discuss the relevance of this result with respect to the properties of the Brillouin zone faces.

CHAPTER 5

Some Applications of Brillouin Zone Theory

THIS chapter will illustrate the use of Brillouin zone theory in the discussion of some properties of metals, in particular phase stability, and the dependence of lattice constants on the electron concentration.

It should be understood that the theories that we shall describe are speculative and not necessarily correct: some are, in fact, demonstrably wrong. A discussion of these theories, however, is a useful methodological example.

1. The Jones theory of the Hume-Rothery rules

In this section and the next two we shall be concerned with the problem of phase stability. We shall first describe some of the features of the experimental situation. Copper, as we know, is f.c.c. As we add zinc to it the f.c.c. phase (α) persists until a certain concentration is reached at which the phase α ceases to be stable and another, b.c.c. phase (β) appears. Hume-Rothery was the first to point out a most remarkable feature of this transition, which relates to the *electron concentration* at which it takes place. In pure copper and pure zinc respectively we can say that the electron concentrations in the Fermi sea are 1 and 2 electrons per atom respectively. As we alloy zinc to copper the electron concentration of the material is increased until a value, say n_α, is reached when the $\alpha \rightarrow \beta$ phase transition takes place. The remarkable fact discovered by Hume-Rothery is that the electron concentration is the determining factor of the transition. In fact, if instead of zinc we alloy copper with other metals such as aluminium or germanium, the value of n_α at which the transition occurs is constant within 2% or 3% around 1·38. Moreover, Hume-Rothery considered numerous other alloy systems such as those in which the copper matrix

200

is replaced by silver and found that they experience an $\alpha \to \beta$ (i.e. face centred to body centred) transition at around the same value $n_\alpha = 1{\cdot}38$. Hume-Rothery also considered other phase transitions and enunciated a set of rules that give the electron concentrations at which they take place. [See Table 1 (p. 202) and discussion following it.]

H. Jones proposed an explanation of the $\alpha \to \beta$ phase transition, which was based on a number of assumptions.

The first assumption entails the acceptance of the so-called *rigid band model*. This provides a very simple picture of an alloy and allows us to deal with it largely as if it were a pure metal. In fact, we assume that as we add zinc to copper the charge around the zinc cores is rearranged in the metal so that they cannot be distinguished from the copper cores. As a result, the band structure of pure copper is left unperturbed (hence the name of rigid band for the model) and all that happens is that the extra electrons added to the system occupy the vacant states (above the Fermi level) in the copper band structure. The Brillouin zone of copper originally half full (one electron per atom only), will gradually fill in as we add more zinc (i.e. more electrons), and the states are occupied by following the undisturbed contour surfaces that correspond to pure copper. The rigid band model appears to be an adequate description for dilute alloys, but it is often applied for high concentrations where its validity is questionable.

The second assumption of the Jones theory is the acceptance of the free-electron model for the systems under consideration, recognizing, however, that energy gaps appear at the Brillouin zone faces. (When this is not the case, as in the h.c.p. lattice, the Jones zone will be used.) Within this assumption, the Fermi surface of copper in the α phase will be a sphere of volume equal to one half that of the f.c.c. Brillouin zone. As we add zinc to copper, the Fermi surface of the α phase expands until an electron concentration is reached, which we shall call ν_α, at which the Fermi surface touches a Brillouin zone face. In Problem 2 of § **4.**17 we saw that this contact occurs at the centre of the hexagonal face of the Brillouin zone (L) at $\nu_\alpha = 1{\cdot}36$. (Notice the remarkable similarity with the concentration $n_\alpha = 1{\cdot}38$ at which the α phase ceases to be stable.)

A third assumption is now made. If for a phase, λ say, the Fermi surface touches the Brillouin zone boundary at a concentration ν_λ, and if at that concentration there is another crystal structure, μ say, for which the Fermi surface is still wholly *inside* the Brillouin zone, it is assumed that μ is more stable than λ and therefore that the transition $\lambda \to \mu$ will take place.

We shall now justify this third and crucial assumption. In fact, after the Brillouin or Jones zone face is touched at the electron concentration ν_λ, further addition of electrons to the λ phase will entail the occupation of states in the second zone. On account of the energy discontinuity at the surface of the zone these extra electrons will experience a sudden rise in energy, thus sharply raising the energy of the whole electron system and thereby reducing its stability with respect to the μ phase for which the Fermi surface is still inside the corresponding Brillouin zone. Moreover, if, as is likely, the Fermi surface of the μ phase is just below the Brillouin zone boundary at the concentration in question, there is an added advantage for the μ phase, since in this region the energy is a slowly varying function of $\mathbf{k}$ (because, at least for some directions on the surface, $(\partial E/\partial \mathbf{k})_n = 0$ at the boundary). This means that, in this region, electrons can be placed in states with higher values of $\mathbf{k}$ without increasing their energy much. (An alternative discussion of this point will be given in § 3.)

TABLE 1. PHASE STABILITY

Phase		ν_λ	n_λ
Face centred cubic	(α)	1·36	1·38
Body centred cubic	(β)	1·48	1·50
Complex	(γ)	1·54	1·62
Hexagonal close packed	(ε)	1·69	1·75

λ takes the values α, β, γ, ε. The values given are electron concentrations in electrons per atom.

In Table 1 we apply the Jones theory and compare it with the Hume-Rothery rules concerning the stability of various alloy phases.

The phases shown in the first column are those that appear in brasses and similar alloys with the Greek letters conventionally used to designate them. The values of ν_λ are the electron concentrations at which the spherical Fermi surface touches the Brillouin (or Jones) zone boundary for the first time, and they are computed as in § **4.17.1** (see also Exercise below). It follows from the table that at the electron concentration 1·36, the spherical Fermi surface touches the f.c.c. Brillouin zone boundary, whereas at that concentration it is still inside the Brillouin zone of the b.c.c., the boundary of which is not touched until $n_\beta = 1\cdot48$. In accordance to our previous discussion, the $\alpha\rightarrow\beta$ transition should take place at the concentration $\nu_\alpha = 1\cdot36$ in remarkable agreement with the experimental value $n_\alpha = 1\cdot38$, given in the last column of the table. The remaining values in this column are the electron concentrations which are given in the *Hume-Rothery rules* as the experimental values at which the corresponding phases are stable. Therefore, they should not be compared directly with the values ν_λ which correspond in the present theory to the electron concentration at which the phase λ ceases to be stable. Also, of course, the values given by the Hume-Rothery rules are rather rough averages of values pertaining to different alloy systems. It is clear, nevertheless, that the present theory predicts correctly the order in which the various phases appear. Moreover, the numerical agreement is quite reasonable. For instance, from the Hume-Rothery rules, the $\beta\rightarrow\gamma$ transition should take place at an electron concentration somewhere between 1·50 and 1·62, to be compared with the theoretical value 1·54. Likewise the $\gamma\rightarrow\varepsilon$ transition should take place between 1·62 and 1·75, the theoretical value being 1·69.

BIBLIOGRAPHICAL NOTE

For the Hume-Rothery rules, see W. Hume-Rothery, *J. Inst. Metals* **35**, 295, 3–7 (1926); *The Metallic State*, Oxford University Press (1931). For the Jones theory see H. Jones, *Proc. Roy. Soc.* A, **144**, 225 (1934); A, **147**, 396 (1934); *Proc. Phys. Soc.* A, **49**, 250 (1937). Also N. F. Mott and H. Jones, *The Theory of the Properties of Metals and Alloys*, Oxford University Press (1936), chapter V.

Exercise

Show that the electron concentration at which the spherical Fermi surface of a b.c.c. structure touches the boundary of the Brillouin zone is 1·48.

Method. The reciprocal lattice of a b.c.c. lattice is f.c.c. (see Exercise 1, § **4.17.1**). This is shown in Fig. 89a. There are $8\frac{1}{8}+6\frac{1}{2}=4$ lattice points in the cubic unit cell shown in the figure. The Brillouin zone, which is a primitive cell, has a volume Ω equal to one quarter of that of the cube shown: $\Omega = \frac{1}{4}d^3$. The Brillouin zone is shown in Fig. 89b. Its centre is the point A in (a) and the hatched faces (and the two parallel to them on the back of the figure) correspond to the bisectors shown around A. The radius of the inscribed sphere is $AB = d/2\sqrt{2}$, and its volume $\Omega' = \pi d^3/12\sqrt{2}$. Verify that $\Omega'/\Omega = 0\cdot74$. Since Ω contains two electrons per atom Ω' contains $2\times0\cdot74 = 1\cdot48$ electrons per atom.

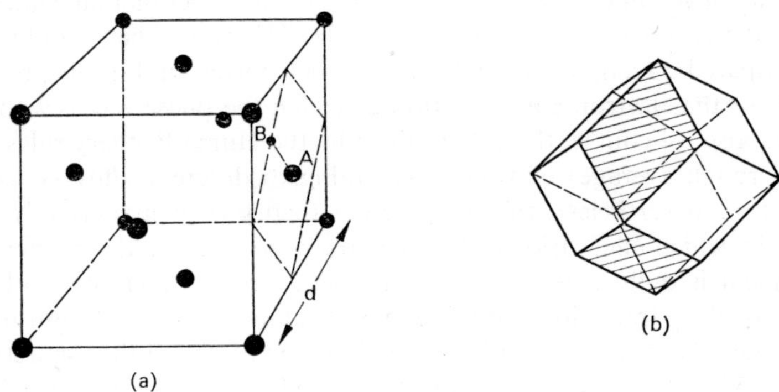

FIG. 89. The b.c.c. lattice Brillouin zone.

2. Further theories of phase stability

Attractive as the Jones theory is on account of its simplicity and of its impressive agreement with experiment, it cannot be right. This is so because its basic assumption is that in f.c.c. copper (electron concentration equal to 1) the Fermi surface is spherical and that as the electron concentration is gradually increased to $1\cdot36$ the Fermi surface makes contact with the hexagonal faces of the Brillouin zone. However, it is now known experimentally that the Fermi surface of copper is distorted enough from the spherical shape as to touch already the hexagonal faces of the Brillouin zone (see § **6.8.3**).

Cohen and Heine suggested the following way in which the simplicity of the Jones theory could be recovered. They argued that when a small amount of zinc is added to copper, the Fermi surface becomes more free-electron like, i.e. more spherical, and detaches itself from the faces of the Brillouin zone. Further additions of zinc causes the Fermi surface to expand until it touches the Brillouin zone at the value that corresponds to the free-electron case, i.e. an electron concentration of 1·36.

Cohen and Heine supported as follows their conjecture that the addition of zinc makes the Fermi surface more free-electron like. At the top of the first band of copper it is known from detailed calculations that the wave functions have essentially p character (just as for the first band in the example of § 3.20), whereas the bottom of the second band is s-like. The spectroscopic study of the atomic electron levels shows that the s level in zinc is much further below the p level than in copper. This would suggest that addition of zinc to copper reduces the gap between the top of the first band and the bottom of the second. If one assumes that the gap is closed, it follows that the bands become free-electron like and that the Fermi surface will become spherical.

Unfortunately, Cohen and Heine's theory is not correct; on comparing the atomic levels of silver and copper the argument used above would predict a spherical Fermi surface for pure silver, whereas it is now known that, as in copper, this touches the hexagonal faces of the Brillouin zone.

In view of this fact Hume-Rothery and Roaf made the following observation. If we accept the correct Fermi surface for copper, in which contact is made with the hexagonal faces, and add electrons to it, the new Fermi surface will expand and eventually touch the more distant square faces. Approximate calculations based on the known form of the Fermi surface indicate that this contact might occur at an electron concentration of about 1·4: Hume-Rothery and Roaf argue that this contact with the square face is the one that determines the $\alpha \rightarrow \beta$ phase change, which happens experimentally at an electron concentration of 1·38 (Table 1).

REFERENCES

M. H. Cohen and V. Heine, *Advances in Physics*, **7**, 395 (1958); W. Hume-Rothery and D. J. Roaf, *Phil. Mag.* **6**, 55 (1961).

3. Peaks in the density of states curves

An argument was given in §1 to show that phase changes are determined by the contact of the Fermi surface with the Brillouin zone boundary. Since this property is often discussed in terms of density of states curves, it will be useful to consider some of their properties. It will first be shown that peaks in the density of states curves appear at the zone boundaries, for which we shall consider Fig. 90.

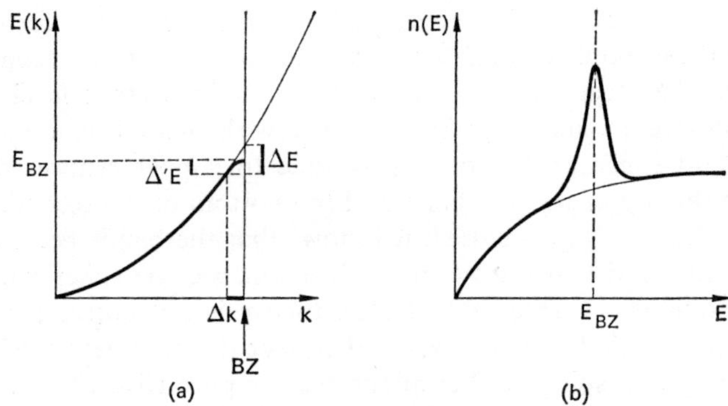

FIG. 90. Peaks in the density of states $n(E)$ curves. *BZ* indicates the value of **k** at the edge of the Brillouin zone and E_{BZ} the corresponding value of the energy for the bound electron case.

The thin lines of the figure correspond to the free electron case: in (a) the free electron parabola $E(k)$ is given and in (b) the corresponding density of states curve $n(E)$ is plotted as in Fig. 18 of Chapter 2. The full lines correspond to properties of the bound electrons. Their $E(k)$ curve departs from that of the free electrons as in Fig. 35 of Chapter 3. Let us see what the effect of this departure is on the change of the density of states curve. The number of free-electron states in the energy range ΔE is directly proportional to the interval Δk of values of k to which they correspond (since the number of states is directly proportional to the volume in k space or length in one dimension). Exactly the same number of states (i.e. the same segment Δk of the k axis) corresponds to the much smaller range of energies $\Delta'E$ for the bound electrons. The number of states per unit of energy will be proportional to $\Delta k/\Delta E$ for the free electrons and to $\Delta k/\Delta'E$ for the bound ones. From Fig. 90, since $\Delta'E$ is about one-half ΔE, the ratio $\Delta k/\Delta'E \approx 2\Delta k/\Delta E$ and the density of states for the bound electrons must be about twice as large as for the free electrons as shown in Fig. 90b. In general, the density of states must show a sharp rise near the edge of the Brillouin zone, whereas after this is passed, since the $E(k)$ curve becomes again similar to the free-electron one (for weak fields), the density of states falls back to the free-electron value.†

We shall now discuss the relevance of the peaks in the density of states curves as regards the problem of phase stability, for which we shall consider the density of states curves for electrons in the f.c.c. and b.c.c. structures. Since the Brillouin zone boundary is touched first in the f.c.c. structure (Table 1) the curves are disposed as in Fig. 91. Let us now consider a f.c.c. alloy for which the electron concentration is such that the contact with the Brillouin zone has been surpassed, so that the Fermi energy is E'. Since we want to compare the state of this alloy in the b.c.c. phase with that in the f.c.c. phase we have chosen for convenience the electron concentration so that E' is also the Fermi energy for the b.c.c. phase. This is not difficult: we remember from (2.44) that the area under the $n(E)$ curve is equal to the total number of electrons, so that we have chosen E' in such a way that the area it limits in either curve is the same. Suppose now that we want to increment the electron concentration by a small amount Δ. We must add an area under the corresponding curve that corresponds to Δ. For the b.c.c. phase this area is given by the shaded strip, whereas for the f.c.c. phase it is the shorter, and therefore broader, hatched strip. It is clear that exactly the same increment in the electron concentration of both phases entails a larger increment of the energy for the f.c.c. phase. This is the relevant feature: because of the sharp decrease in the density of states curve after the Brillouin zone edge is touched, an increment in the

† ●●●The reader who has followed Exercise 2 of § 4.14 will be able to derive the above result by noticing that, on account of the properties discussed in § 4.13, $|\text{grad}_k E| = 0$ either over the whole of the Brillouin zone face or along certain directions on it. In the second case, even if the gradient does not vanish it becomes very small over the whole face, which produces a very large increase in the integral that defines $n(E)$.

electron concentration is energetically very costly. If, instead, we add electrons in a region where the density of states is rising sharply an increase of the electron concentration can be achieved with hardly a change in the total energy.

The above argument shows that the electron concentration at which the Brillouin zone boundary is touched is a critical one, after which the addition of further electrons is energetically expensive, so that another

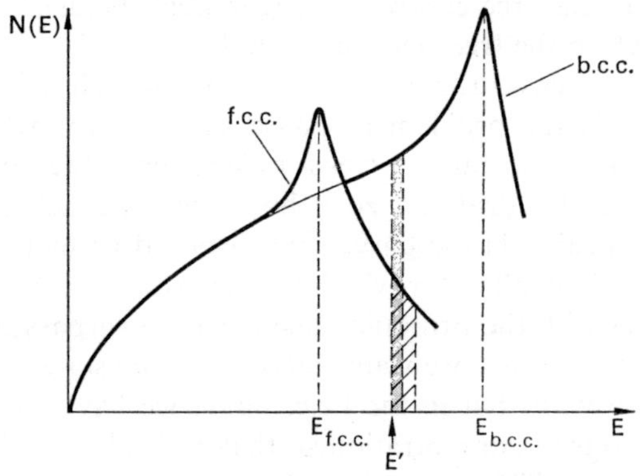

FIG. 91. Density of states curves and phase stability. The energy values $E_{f.c.c.}$ and $E_{b.c.c.}$ are those at which contact is established with the boundaries of the f.c.c. and b.c.c. Brillouin zones respectively.

structure for which contact with the Brillouin zone has not yet been effected will be more stable. This is essentially the concept used in the last section, although it was justified there in a different way.

It should not be thought that all peaks in the density of states curves are necessarily associated with contacts of the Fermi surface with the Brillouin zone boundary. For example, consider the bands in Fig. 92. Exactly the same number of states belong to the energy interval ΔE_2 as to ΔE_1 (since the same region of **k** space pertains to both). Since $\Delta E_2 \ll \Delta E_1$, the density of states at values of the energy in the region ΔE_2 will be much larger than for those in ΔE_1. This, of course, is quite general: a narrow band determines a peak in the density of states curve.

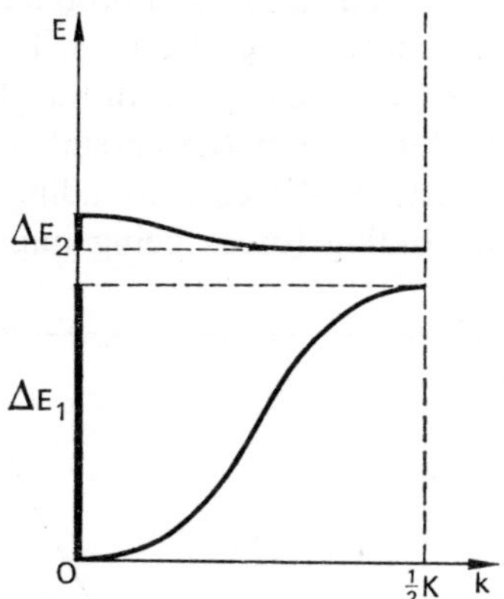

FIG. 92. Narrow bands correspond to a
high density of states.

4. Effect of the Fermi energy on lattice parameters

The free-electron Fermi energy is proportional to the electron concentration [see (2.39)] and it is clear that this should be so also in the
general case if the rigid band picture holds. Since the electron concentration decreases when the lattice constant of a crystal expands, expansion
of the lattice will produce a decrease of the Fermi energy. A crystal, in
general, will try and decrease its Fermi energy but there is a limit to the
extent it can achieve this by expanding; this is so because, on expansion,
stability is gained by a reduction in Fermi energy but it is lost by a
weakening of the binding forces of the crystal. It is this second counter-
acting effect that prevents the crystal from blowing up to infinity in an
attempt to reduce its Fermi energy.

When the electron concentration is small the Fermi energy is not
large and the determining factor in the energy is given by the binding
forces; in this situation, as the electron concentration increases, the
stronger becomes the binding and the lattice constant diminishes therefore without regard to the corresponding increase in the Fermi energy
which, being small, is comparatively unimportant.

On the other hand, when the electron concentration is large enough
for the Fermi surface to touch the Brillouin zone boundary, a small

increase in the electron concentration will produce a very large increase in the Fermi energy, as shown in §§ 1 and 3. From this moment on, the sharp increase in the Fermi energy with the electron concentration becomes the dominant factor and the crystal must counteract it by increasing its lattice constant, the corresponding increase of the now relatively unimportant binding forces being immaterial.

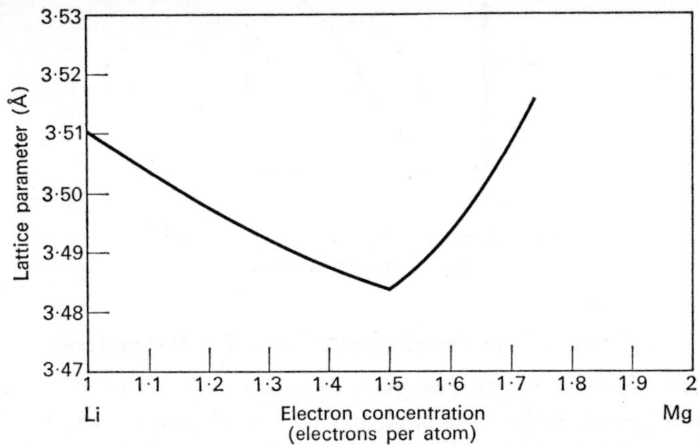

FIG. 93. Variation of the lattice parameter of lithium–magnesium alloy with the electron concentration. [After D. W. Levinson, *Acta Met.* **3**, 294 (1955).]

An extremely good example of the situation just described is provided by the b.c.c. alloy of lithium–magnesium. From Table 1, we know that the Fermi surface touches the b.c.c. Brillouin zone boundary at an electron concentration of 1·48 electrons per atom, so that as magnesium is added to lithium we should expect the lattice constant to decrease at first until a concentration about 1·5 is reached, when the Fermi energy becomes a dominant factor so that further increases in the electron concentration would cause an expansion of the lattice parameter. This is exactly what the experiments show (Fig. 93). The reader should appreciate, nevertheless, that the experimental interpretation given provides in general no more than semi-quantitative guidance and that the astonishing agreement between theory and experiment shown in the figure should not be over-emphasized.

CHAPTER 6

The Calculation of Band Structures and Fermi Surfaces

1. The problem and the basic equations

We want to obtain solutions of the Schrödinger equation for the N electron Fermi sea,

$$\mathbf{H}\psi = E\psi. \tag{1}$$

As we have seen in Chapter 2 [see (2.17)] the solution of (1) is drastic-ally simplified by the one-electron approximation, whereby $\mathbf{H}$ is written as a sum of effective hamiltonians $^e\mathbf{H}$ for each particle, in which no interaction terms appear. The crucial feature of these effective hamiltonians is that they contain an averaged potential in which the particles move. The only way in which the effective potentials pertaining to different particles differ is through the contribution due to the particle for which the Schrödinger equation is being solved. On account of the large number of particles, the contribution of any one of them to the crystal potential is very small so that all averaged potentials are taken to be the same and the N effective Schrödinger equations reduce to a single one

$$^e\mathbf{H}\varphi_{\mathbf{k}}^{j}(\mathbf{r}) = E_{\mathbf{k}}^{j}\varphi_{\mathbf{k}}^{j}(\mathbf{r}) \tag{2}$$

in which, as shown in Chapter 3, we recognize that the solutions must be Bloch functions

$$\varphi_{\mathbf{k}}^{j}(\mathbf{r}) = \exp\left(2\pi i\mathbf{k}.\mathbf{r}\right)u_{\mathbf{k}}^{j}(\mathbf{r}). \tag{3}$$

In (2) and (3) $\mathbf{k}$, which takes all values over the Brillouin zone, and j which denotes a band, are indices that label the different one-electron eigenfunctions. In order to construct the *total* eigenfunctions ψ of the

211

H

N electron Fermi sea given by (1), we must obtain the N solutions of (2) that correspond to the N lowest energy levels and multiply them through (see Chapter 2).

We must consider briefly the form of the hamiltonian in (2) for which we shall henceforth drop the superscript e. The one-dimensional hamiltonian of (1.22) can be extended very readily to three dimensions as follows:

$$\mathbf{H} = -\frac{h^2}{8\pi^2 \mathbf{m}}\left(\frac{d^2}{dx^2}+\frac{d^2}{dy^2}+\frac{d^2}{dz^2}\right)+V(xyz), \qquad (4)$$

where $V(xyz)$ is the effective potential and the operator in brackets is called the *Laplacian* ∇^2. $V(xyz)$ has the full crystal periodicity and it is further simplified by the use of the *spherical approximation*, which will require a slight change of coordinates, illustrated in Fig. 94. We divide up the crystal lattice in centred unit cells (first introduced by Wigner and Seitz in this connection and often called Wigner–Seitz cells). The nth cell is denoted by the vector of the lattice $\mathbf{R}_n$ (see § 4.1). The coordinates

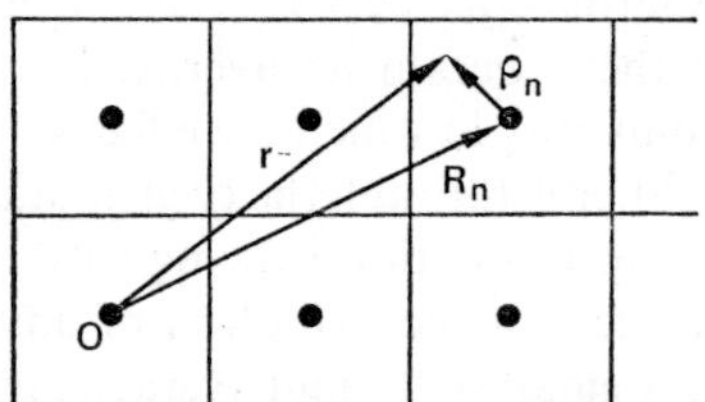

FIG. 94. Position vectors in the crystal lattice.
The point O is the origin of the lattice.

xyz in (4) denote the vector $\mathbf{r}$ which we want to replace by the vector $\boldsymbol{\rho}_n$ referred to the origin of the nth cell

$$\boldsymbol{\rho}_n = \mathbf{r} - \mathbf{R}_n. \qquad (5)$$

The spherical approximation assumes that within each cell the potential is only a function of ρ, that is of its distance to the central atom in the cell. Therefore within the nth cell $V(xyz)$ can be written as $V(\rho_n)$ and on account of the periodicity condition $V(\boldsymbol{\rho}_n) = V(\boldsymbol{\rho}_m)$.

From the above, we can write (4) as follows:

$$\mathbf{H} = -\frac{h^2}{8\pi^2\mathbf{m}}\nabla^2 + V(\rho). \tag{6}$$

Although $\mathbf{k}$ does not appear explicitly in the hamiltonian (6) it is highly relevant in determining its eigenvalues since it appears in the boundary conditions that must be satisfied by the eigenfunctions (3):

$$\mathbf{R}_\nu\varphi_\mathbf{k}^j(\mathbf{r}) = \exp(2\pi i\mathbf{k}.\mathbf{R}_\nu)\varphi_\mathbf{k}^j(\mathbf{r}). \tag{7}$$

In the following sections we shall discuss several methods to obtain an approximate solution of (2).

● 2. The tight-binding method

1. THE BLOCH SUMS

The tight-binding method was the first technique, proposed by F. Bloch,[†] to obtain the band structure of a crystal. The basic idea is this: instead of solving (2) directly one tries to form approximate eigenfunctions that satisfy the fundamental Bloch condition (7). If the electrons were tightly bound to the nuclei (as in the case of large inter-nuclear separation) the wave function (3) would coincide, within the nth cell, with an atomic eigenfunction ϕ_n^j. The suffix here denotes the cell to which the function pertains and the superscript denotes one of the atomic eigenfunctions (that is for a lattice of hydrogen atoms, say, one of the functions $1s$, $2s$, $2p$, etc.). Since the functions $\phi_n^j(n = 1, 2, \ldots, N)$ are adequate forms of the wave function in the limiting case considered, we expect that a linear combination of them

$$\varphi_\mathbf{k}^j(\mathbf{r}) = \sum_1^N c_n\phi_n^j \tag{8}$$

will be a good approximation to the wave function. (This type of approximation, which is based on the variational scheme of § **1.18** is now very much used in the theory of molecular structure, under the name of

[†] *Z. f. Phys.* **52**, 555 (1928).

LCAO, or *linear combination of atomic orbitals method*.) In the general case the coefficients c_n have to be obtained by the solution of the variational equations (see § **1**.18). Bloch noticed, however, that the functions (8) have to satisfy the fundamental condition (7) and showed that as a result of this the coefficients c_n are fully determined, a very powerful result indeed, which we shall now demonstrate.

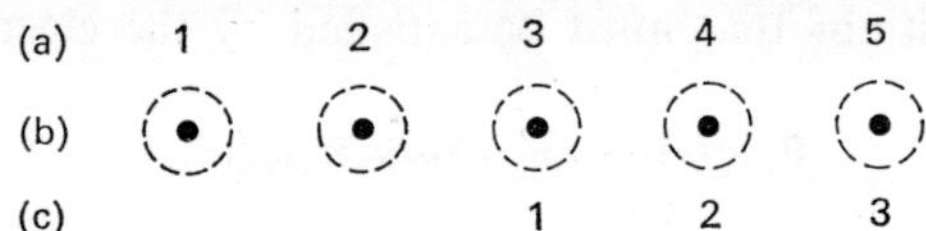

FIG. 95. Effect of translations on the orbitals ϕ_n. In (b), the circles represent the orbitals ϕ_n, originally labelled as in (a). In (c), a translation $\mathbf{R}_2$ has been effected, which causes the re-labelling of the orbitals shown. It follows that $\mathbf{R}_2 \phi_4$, say (which means the function ϕ_4 after $\mathbf{R}_2$ has been performed), is ϕ_2, or, more generally, that $\mathbf{R}_2 \phi_n = \phi_{n-2}$. For a general translation $\mathbf{R}_\nu$, equation (10) of the text follows.

Operate on both sides of (8) with $\mathbf{R}_\nu$:

$$\mathbf{R}_\nu \varphi_{\mathbf{k}}^j(\mathbf{r}) = \sum_{1}^{N} c_n \mathbf{R}_\nu \phi_n^j. \tag{9}$$

A little attention has to be paid to the meaning of the operator relation on the right-hand side of (9). We show in Fig. 95 that

$$\mathbf{R}_\nu \phi_n^j \equiv \phi_{n-\nu}^j, \tag{10}$$

whence

$$\mathbf{R}_\nu \varphi_{\mathbf{k}}^j(\mathbf{r}) = \sum_{1}^{N} c_n \phi_{n-\nu}^j. \tag{11}$$

It can be seen at once that, in order to satisfy (7),

$$c_n = \exp(2\pi i \mathbf{k}.\mathbf{R}_\nu) c_{n-\nu}. \tag{12}$$

In fact, with (12) in (11),

$$\mathbf{R}_\nu \varphi_{\mathbf{k}}^j(\mathbf{r}) = \exp(2\pi i \mathbf{k}.\mathbf{R}_\nu) \sum_{1}^{N} c_{n-\nu} \phi_{n-\nu}^j \tag{13}$$

and on account of periodicity the summations on the right-hand sides of (13) and (8) are identical,† whence (7) is satisfied.

From (12) it follows that c_n must depend exponentially on n, and it is easy to guess the details of the exponent so as to satisfy that relation:

$$c_n = \exp(2\pi i \mathbf{k}.\mathbf{R}_n). \tag{14}$$

In fact, from (14) for the suffix $n-\nu$:

$$c_{n-\nu} = \exp(2\pi i \mathbf{k}.\mathbf{R}_{n-\nu}) \tag{15}$$

$$= \exp[2\pi i \mathbf{k}.(\mathbf{R}_n - \mathbf{R}_\nu)] \tag{16}$$

$$= \exp(-2\pi i \mathbf{k}.\mathbf{R}_\nu)c_n, \tag{17}$$

which satisfies (12).

On introducing (14) in (8) we obtain the final form of the approximate Bloch functions:

$$\varphi_\mathbf{k}^j(\mathbf{r}) = \sum_1^N \exp(2\pi i \mathbf{k}.\mathbf{R}_n)\phi_n^j. \tag{18}$$

Because of their form, these functions are often called *Bloch sums.* It should be noticed that each atomic level will generate one set of Bloch sums, since in (18) $\mathbf{k}$ can take all N values in the first Brillouin zone.

1. *Remark*

It is easy to write the Bloch sums (18) in the more general form (3). In fact, from (18),

$$\varphi_\mathbf{k}^j(\mathbf{r}) = \exp(2\pi i \mathbf{k}.\mathbf{r}) \sum_1^N \exp[2\pi i \mathbf{k}.(-\mathbf{r}+\mathbf{R}_n)]\phi_n^j, \tag{19}$$

and, from (5),

$$\varphi_\mathbf{k}^j(\mathbf{r}) = \exp(2\pi i \mathbf{k}.\mathbf{r}) \sum_1^N \exp(-2\pi i \mathbf{k}.\boldsymbol{\rho}_n)\phi_n^j. \tag{20}$$

---

† The functions that appear in (8) are $\phi_1, \phi_2, \ldots, \phi_N$. Those in (13) are $\phi_{1-\nu}$, $\phi_{2-\nu}, \ldots, \phi_0, \phi_1, \ldots \phi_{N-\nu}$. From periodicity $\phi_{1-\nu} = \phi_{N+(1-\nu)} = \phi_{N-\nu+1}$, $\phi_{2-\nu} = \phi_{N-\nu+2}, \ldots, \phi_0 = \phi_{\nu-\nu} = \phi_{N-\nu+\nu} \equiv \phi_N$, so that the second sequence is identical to the first. The coefficients must follow suit, since they are attached to the functions that they multiply.

It can readily be proved that the summation here has the periodicity property required of the cell function $u_k^j(\mathbf{r})$ in (3).

2. THE ENERGY EIGENVALUES

Since the coefficients c_n in (8) have been fully determined from the Bloch condition in (18) there is no need of a variational calculation to obtain them (but see § 2.2.1). It is enough, in order to obtain the energy that corresponds to φ_k^j, to substitute it for f in (1.74) since φ_k^j contains no variable parameters:

$$E_k^j = \int (\varphi_k^j)^* \mathbf{H} \varphi_k^j d\tau \Big/ \int (\varphi_k^j)^* \varphi_k^j d\tau. \tag{21}$$

How this is done will be clear in the exercise of § 2.3.

1. *The secular determinant* ($\rightarrow$ *end*)

If we had taken the variational expression (8) straightaway we would have had to solve the secular determinant (1.78) of order $N \times N$, a huge if not an impossible task. Instead, we require for each φ_k^j the single term (21) so that the simplification achieved is extraordinary. We want to look into this in more detail. In principle we have replaced the N functions $\phi_n^j (n = 1, 2, \ldots, N)$ in (8) by the N combinations (18) φ_k^j (k over the N values in the first Brillouin zone). The treatment that we have given so far is therefore an oversimplification. We must recognize that we ought to replace the linear combination (18) by a similar linear combination of the new functions φ_k^j

$$\psi(\mathbf{r}) = \sum_{\text{all } k} c_k \varphi_k^j. \tag{22}$$

This would lead to an $N \times N$ secular determinant

$$|H_{kk'} - ES_{kk'}| = 0 \tag{23}$$

with

$$H_{kk'} = \int (\varphi_k^j)^* \mathbf{H} \varphi_{k'}^j \, d\tau, \tag{24}$$

$$S_{kk'} = \int (\varphi_k^j)^* \varphi_{k'}^j \, d\tau. \tag{25}$$

A treatment similar to that of § **3.22.1.6** shows that the matrix elements vanish if $\mathbf{k} \neq \mathbf{k}'$. Take the so-called *overlap integral* $S_{\mathbf{k}\mathbf{k}'}$ for instance. This cannot be altered by the symmetry operation $\mathbf{R}$:

$$S_{\mathbf{k}\mathbf{k}'} = \mathbf{R}S_{\mathbf{k}\mathbf{k}'} = \int (\mathbf{R}\varphi_{\mathbf{k}}^{j})^* \mathbf{R}\varphi_{\mathbf{k}'}^{j}\, d\tau$$

$$= \exp\left[2\pi i(\mathbf{k}' - \mathbf{k}).\mathbf{R}\right] \int (\varphi_{\mathbf{k}}^{j})^* \varphi_{\mathbf{k}'}\, d\tau. \tag{26}$$

[In the last step we use (7).] Since the exponential in (26) is different from unity for $\mathbf{k}' \neq \mathbf{k}$ the integral must vanish when this is not the case. The corresponding result for $H_{\mathbf{k}\mathbf{k}'}$ is proved similarly. It follows, as in § **4.22.1**, that the secular determinant factorizes out in blocks, each block corresponding to one value of $\mathbf{k}$. In the linear chain there is only one function that belongs to each $\mathbf{k}$ value, so that each block will have only one term and (21) follows. In three-dimensional crystals or even in linear lattices with bases it is possible, for some values of $\mathbf{k}$, to have m functions $(m > 1)$ corresponding to a given $\mathbf{k}$, and this will lead to secular determinants of order $m \times m$.

3. Exercise: Tight-binding treatment of a linear chain

To obtain the Bloch functions and the corresponding eigenvalues for a linear chain of six atoms with periodic boundary conditions [see Figs. 26 and 27 (pp. 71 and 72)].

1. *Formulae*

Since we have six atoms there must be six values of $\mathbf{k}$ in the Brillouin zone. From (3.26) these are

$$\mathbf{k} = \frac{\kappa}{6a}, \quad \kappa = -3, -2, -1, 0, 1, 2, \tag{27}$$

where a is the lattice constant. For simplicity, we shall replace $\mathbf{k}$ by κ as a label.

In order to write the Bloch sums from (18) we notice that $\mathbf{R}_n = n\mathbf{a}$, so that, from (27) $\mathbf{k}.\mathbf{R}_n = \kappa n/6$. The Bloch sums take the form

$$\varphi_\kappa = \sum_{n=1}^{6} \exp\left(\frac{2\pi i}{6}\,\kappa n\right)\phi_n, \qquad \kappa = -3, -2, \ldots, 2, \tag{28}$$

where for the moment we dispense with the band label j.

2. *The Bloch sums*

Equation (28) can be written as

$$\varphi_\kappa = \sum_{n=1}^{6} (\omega^\kappa)^n \phi_n, \quad \kappa = -3, \quad -2, \quad -1, \quad 0, \quad 1, \quad 2, \tag{29}$$

with

$$\omega = \exp(2\pi i/6). \tag{30}$$

The values of ω^κ are shown in Fig. 96 in the usual way for complex numbers. From this figure the various powers of ω^κ are quickly obtained, as displayed in Table 1.

TABLE 1. VALUES OF $(\omega^\kappa)^n$, $\omega = \exp(2\pi i/6)$

κ \ n	1	2	3	4	5	6
0	1	1	1	1	1	1
1	ω	$-\omega^*$	-1	$-\omega$	ω^*	1
2	$-\omega^*$	$-\omega$	1	$-\omega^*$	$-\omega$	1
3	-1	1	-1	1	-1	1
-2	$-\omega$	$-\omega^*$	1	$-\omega$	$-\omega^*$	1
-1	ω^*	$-\omega$	-1	$-\omega^*$	ω	1

The Bloch functions (29) can be written at once from Table 1 by reading the coefficients from each line. For example,

$$\varphi_0 = \varphi_1 + \varphi_2 + \varphi_3 + \varphi_4 + \varphi_5 + \varphi_6, \tag{31}$$

$$\varphi_1 = \omega\varphi_1 - \omega^*\varphi_2 - \varphi_3 - \omega\varphi_4 + \omega^*\varphi_5 + \varphi_6. \tag{32}$$

Form in the same manner the other four Bloch sums.

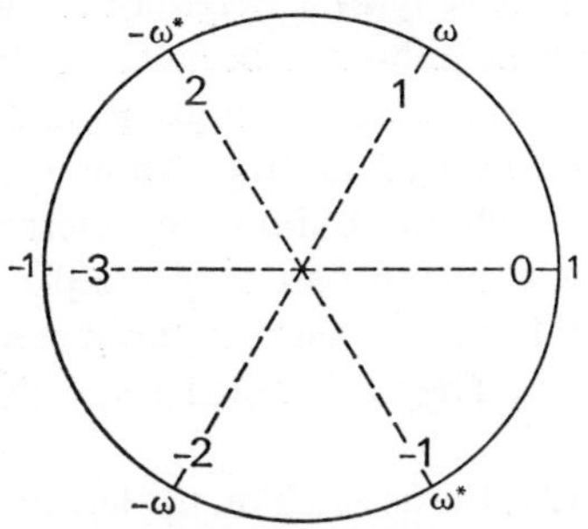

FIG. 96. The values of ω^κ for $\omega = \exp(2\pi i/6)$. The values of κ appear inside the circle and those of ω^κ outside it. $\omega^\kappa = \omega^{\kappa+6}$, whence $\omega^{-3} = \omega^3$, $\omega^{-2} = \omega^4$, $\omega^{-1} = \omega^5$.

3. *Computation of the energy*

When φ_κ, given by expressions like (31) and (32), is introduced in (21) the integrals therein will split into a number of integrals over the atomic orbitals ϕ_n.

In dealing with these integrals it is customary to introduce the following approximations:

$$\int \phi_n \mathbf{H} \phi_n d\tau = \alpha \quad \text{(constant for all } n\text{);} \tag{33}$$

$$\int \phi_n \mathbf{H} \phi_m d\tau = \beta \quad (n \text{ and } m \text{ contiguous}),$$
$$= 0 \quad \text{(otherwise);} \tag{34}$$

$$\int \phi_n \phi_m d\tau = 1 \quad (n = m),$$
$$= 0 \quad (n \neq m). \tag{35}$$

(33) is tantamount to assuming that all the ϕ_n are physically equivalent, which will be valid for a long chain except for a negligible small number of atoms near the boundary. (Not so good an approximation for a linear chain of six atoms, of course.) Equation (34) is a simplifying assumption

H*

equivalent to saying that only first neighbours interact. Care must be exercised when finding neighbours: because of the periodic boundary conditions atoms 1 and 6 are neighbours [cf. Figs. 26 and 27 (pp. 71, 72)]. The first equation in (35) states that the atomic orbitals are normalized. The second is a very rough approximation, referred to as the *neglect of overlap*. The integrals α and β can in principle be computed but in practice they are often left as adjustable parameters which can be fitted from experimental data. They are called respectively the *coulomb* and *exchange* integrals.

When dealing with (21) the asterisk on the left should not be forgotten. For example, the first two terms in the numerator corresponding to (32) will be

$$\int (\omega\phi_1)^* \mathbf{H} \omega\varphi_1 d\tau = \int \phi_1 \mathbf{H} \phi_1 d\tau = \alpha, \tag{36}$$

$$\int (\omega\phi_1)^* \mathbf{H}(-\omega^*\phi_2)d\tau = \omega \int \phi_1 \mathbf{H} \phi_2 d\tau = \omega\beta. \tag{37}$$

Here we recognize that the atomic orbitals are real and that $\omega^*\omega = 1$, $(\omega^*)^2 = -\omega$ (see Fig. 96). It will also be useful to notice that $\omega + \omega^* = 1$. In this manner you will find $E_1 = (6\alpha + 6\beta)/6$. The other eigenvalues are likewise obtained and are displayed in Table 2.

TABLE 2. EIGENVALUES FOR THE LINEAR CHAIN

κ	E_κ	$E_{reduced}$	$E_{free\ electrons}$
0	$\alpha + 2\beta$	0	0
± 1	$\alpha + \beta$	1	1
± 2	$\alpha - \beta$	3	4
-3	$\alpha - 2\beta$	4	9

If we assume β to be negative E_0 will be the lowest state. In the third column we take this state to be the zero of energy and the other eigenvalues are referred to it in units of β for simplicity. In the last column we give the free electron eigenvalues $h^2k^2/8\pi^2\mathbf{m}$ which are proportional to κ^2 in some appropriate units. Notice that the results of Table 2 show the symmetry $E(k) = E(-k)$ as expected. The values of $E(k)$ or rather E_κ are plotted in the first Brillouin zone in Fig. 97 where for comparison the free-electron results are also shown. In order to display the symmetry better we add the extra value $\kappa = 3$, i.e. the last point on the edge of the

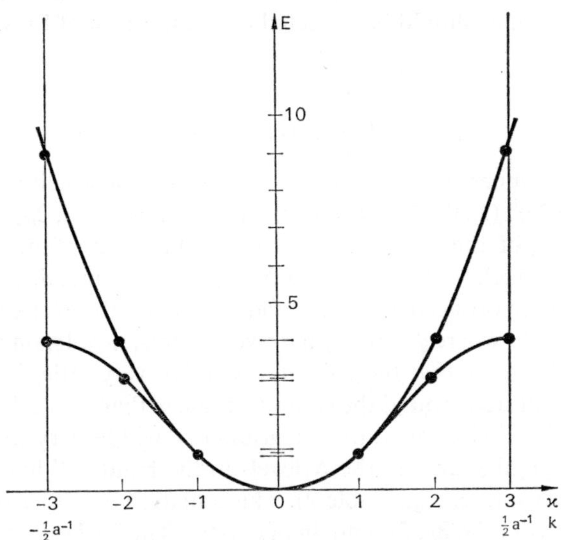

FIG. 97. Energy eigenvalues for the linear chain of six atoms. The upper curve corresponds to free electrons and the lower one to the linear chain, in units of β. The energy eigenvalues for the chain are marked with long lines on the energy scale, a double line denoting a degenerate level. The values of **k** are obtained from (27) and mark the edges of the Brillouin zone. [Cf. Fig. 35 (p. 91), where $K = a^{-1}$.] Notice that a single atomic level has broadened into six for the lattice, as predicted in Fig. 29 (p. 80) for $N = 6$.

Brillouin zone, since for large N this would not alter significantly the number of states. The curves shown in the figure are, of course, not significant, but only the points marked on them. Nevertheless, it can be seen that the curves show the general type of behaviour discussed in Chapter 3.

If $\beta > 0$, E_{-3} is the lowest eigenvalue: the energy has a minimum at the edge of the Brillouin zone and a maximum at its centre. This corresponds to the behaviour of the second band in Fig. 37 (p. 96).

It is interesting to notice that the energy levels and eigenfunctions obtained are exactly those for the benzene molecule in the LCAO approximation, as should be expected on comparison of Figs. 26 and 27 (pp. 71, 72).

4. s AND d BANDS

Consider copper, the electronic configuration of which is $(1s)^2$ $(2s)^2$ $(2p)^6$ $(3s)^2$ $(3p)^6$ $(3d)^{10}$ $(4s)$. For each of the levels $1s$, $2s$, $2p_x$, etc., of the configuration of the free atom a tight binding wave function can be constructed which will give rise to N levels in the metal, forming one band. The main properties of this band depend on the exchange integral $\beta = \int \phi_1 H \phi_2 d\tau$, where ϕ_1 and ϕ_2 are two orbitals, says $1s$, on neighbouring atoms. Suppose that these orbitals overlap very little, i.e. that they are so concentrated around the respective nuclei that where ϕ_1 is large ϕ_2 is very small and vice versa: the integrand in β will be very small indeed, β will be negligible and all the N levels in the band will have the same value, namely α (see, e.g., Table 2). This situation is in fact the case for the core levels $1s$, $2s$, $2p$, $3s$, and $3p$ of copper. The $3d$ levels, on the other hand, overlap to an appreciable extent although a great deal less than the $4s$ levels, which are very spread. It follows therefore that the five $3d$ levels will give rise to five narrow bands, all very near in energy since the free-atom levels are degenerate. Each of these bands will be occupied by two electrons. The $4s$ band, on the other hand, will be very wide and free-electron like. The $3d$ and $4s$ levels are very near in energy in the free atom: although the latter is the higher in energy in this case, in the metal, the bottom of the wide $4s$ band can be lower in energy than that of the $3d$ band. This is in fact the case, that is, the two bands overlap. This can be seen in Fig. 104 (p. 239), where the density of states of copper is shown. This curve can be understood as the superposition of a flat, free-electron like $4s$ band to which a narrow peak is added, corresponding to the $3d$ band. This situation is typical of the transition metals, except that, for them, the $3d$ band will carry less than its full quota of 10 electrons, and the Fermi energy will cut it across.

● 3. The nearly free-electron and orthogonalized plane waves method

We have discussed in § 4.22 the basic principle of the nearly free-electron approximation (NFE). We expand the desired function $\Psi_\mathbf{k}$ in terms of all the plane waves $\psi_{\mathbf{k}+\mathbf{K}_\nu}$ that pertain to the same $\mathbf{k}$

$$\psi_{\mathbf{k}+\mathbf{K}_\nu} = \exp\left[2\pi i(\mathbf{k}+\mathbf{K}_\nu).\mathbf{r}\right], \tag{38}$$

where $\mathbf{K}_\nu$ is a vector of the reciprocal lattice. The expansion takes the form

$$\Psi_\mathbf{k} = \sum_\nu c_\nu \psi_{\mathbf{k}+\mathbf{K}_\nu}, \tag{39}$$

with a slight change of notation from **4**.102. Equation (39) is an infinite series, for all values of $\mathbf{K}_\nu$, and it is found in practice that a very large number of terms is required to obtain a reasonably accurate wave function. The reason for this is the following. Near the ion cores the wave function $\Psi_\mathbf{k}$ must look very much like atomic core functions which oscillate very rapidly, that is which have a large number of nodes in a very small region of space: in order to represent them with plane waves high frequencies are required, that is very large values of $\mathbf{K}_\nu$. It follows that (39) is a slowly convergent series. There is another disadvantage associated with its use, which we shall now discuss.

Imagine that we construct tight-binding wave functions $\varphi_\mathbf{k}^j$ (18) for the core levels. (In fact the tight-binding approximation should be very good for them since if ϕ^j in (18) is one of the core levels, it should be a very good approximation to the wave function near the nucleus.) Since $\Psi_\mathbf{k}$ and $\varphi_\mathbf{k}^j$ belong to the same value of $\mathbf{k}$ they are not necessarily orthogonal, as follows from the result obtained from (26). (The reader can modify the notation in this equation to adapt it to the present case.) We saw in Exercise 3 of § 1.19 that two non-orthogonal functions are not independent. This means that whether we like it or not (39) incorporates a certain amount of the core function $\varphi_\mathbf{k}^j$. This is the source of the trouble: when we perform a variational calculation on a wave function we push down the corresponding energy as much as possible, keeping it always above the energy level we wish to compute. This, for (39), is an energy level of the Fermi sea, not one of the core states. Nevertheless, when minimizing the energy with (39) we are bound to move towards the energy of a core level, since (39) represents partly this level as well. This can introduce large inaccuracies in the eigenvalues. We avoid this difficulty as follows. Instead of using the functions $\psi_{\mathbf{k}+\mathbf{K}_\nu}$ in (39) we shall introduce new functions $\psi_{\mathbf{k}\nu}$ which we shall make orthogonal to all the Bloch sums $\varphi_{\mathbf{k}}^j$ that correspond to all the core levels $j = 1, 2, \ldots$ These functions are called *orthogonal plane waves* (OPW):

$$\psi_{\mathbf{k}\nu} = \psi_{\mathbf{k}+\mathbf{K}_\nu} - \sum_j \lambda_{j\nu}\varphi_{\mathbf{k}}^j. \tag{40}$$

We must choose the λ's so that

$$\int (\varphi_{\mathbf{k}}^{j'})^* \psi_{\mathbf{k}\nu} d\tau = 0 \quad \text{for all} \quad j'. \tag{41}$$

In fact, multiply both sides of (40) with $(\varphi_{\mathbf{k}}^{j'})^*$, integrate and use (41). Since the core functions can be assumed orthogonal $\int (\varphi_{\mathbf{k}}^{j'})^*\varphi_{\mathbf{k}}^j d\tau = \delta_{jj'}$, the right-hand side of the equation gives

$$\int (\varphi_{\mathbf{k}}^{j'})^* \psi_{\mathbf{k}+\mathbf{K}_\nu} d\tau - \lambda_{j'\nu} = 0, \tag{42}$$

which is the condition that λ must satisfy in order to verify the orthogonality relation (41). We can now rewrite the functions $\psi_{\mathbf{k}\nu}$ of (40)

$$\psi_{\mathbf{k}\nu} = \psi_{\mathbf{k}+\mathbf{K}_\nu} - \sum_j \left[\int (\varphi_{\mathbf{k}}^j)^* \psi_{\mathbf{k}+\mathbf{K}_\nu} d\tau \right] \varphi_{\mathbf{k}}^j, \tag{43}$$

and instead of (39) we shall use the variational function

$$\Psi_{\mathbf{k}} = \sum_\nu c_\nu \psi_{\mathbf{k}\nu}, \tag{44}$$

which is correctly orthogonalized to the core levels. An added advantage of (44) is that since the core levels appear in (43) there is no need to use large values of $\mathbf{K}_\nu$ in the expansion. Indeed, the convergence of (44) is often excellent.

The NFE approximation is discussed at length in the book by Mott and Jones mentioned in the General References. The OPW method was introduced by C. Herring† in 1940.

† C. Herring and A. G. Hill, *Phys. Rev.* **58**, 132 (1940).

4. The cellular method

Let us consider one of the centred unit cells, or Wigner–Seitz cells, of Fig. 94, now reproduced in Fig. 98. Within the spherical approximation for the potential the Schrödinger equation (2) must be solved in this cell with the hamiltonian (6), where the potential $V(\rho)$ is only a function of

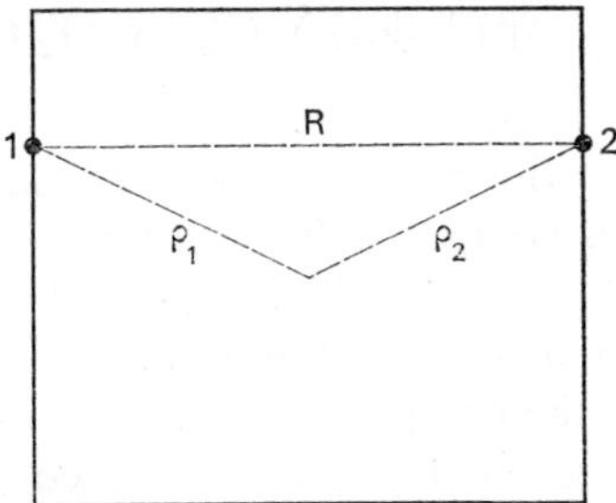

FIG. 98. Boundary conditions in the
cellular method.

the radius ρ measured from the centre of the cell. In order to build up an approximate solution for the crystal, let us assume that the unit cells are at infinite separation: the problem becomes then an atomic problem and the solutions of (2) must go in this limit into atomic functions ϕ^j:

$$\mathbf{H}\phi^j = E^j\phi^j. \tag{45}$$

It is well known from atomic theory that in this case the functions ϕ^j can be written as follows:

$$\phi^j = R_l(\rho)\, Y_l^m(\theta, \phi), \tag{46}$$

where the functions $Y_l^m(\theta, \phi)$ are *spherical harmonics* that depend on the spherical polar angles θ (co-latitude) and ϕ (azimuth) and the *radial functions* $R_l(\rho)$ are solutions of the *radial Schrödinger equation*

$$R_l'' + \frac{2}{\rho}\, R_l' + \left[2V(\rho) + \varepsilon - \frac{l(l+1)}{\rho^2} \right] R_l = 0, \tag{47}$$

where the energy parameter ε, on minimization, will take the value of the eigenvalue E^j in (45) in rydberg units. It should be noticed that the label j of the atomic wave function and eigenvalue stands for the two quantum numbers l (*azimuthal*) and m (*magnetic*).

Since the functions (46) are correct solutions of our problem for infinite internuclear distances, it is reasonable to express the solution for finite separation as

$$\varphi_\mathbf{k}^j = \exp\left(2\pi i \mathbf{k}.\mathbf{r}\right) \sum_{lm} C_{lm}^{kj} R_l(\rho) \, Y_l^m(\theta\phi), \tag{48}$$

where $\mathbf{r}$ is the radius to the crystal origin and ρ that to the centre of the unit cell at which $\varphi_\mathbf{k}^j$ is computed. The exponential in (48) is added, of course, to satisfy the Bloch form (3). The problem reduces to finding the coefficients C_{lm}^{kj} which, as the notation implies, depend on the $\mathbf{k}$ vector and band j for which the energy is computed. This is done as follows. First, the infinite expansion (48) is cut off at a convenient value of l for which the number of unknowns is n. (Usually $l = 6$, leading to 49 coefficients or, since these are in general complex, to $n = 98$ unknowns). Secondly, *boundary condition equations* are set up in order to determine the coefficients. In fact, from (7), it follows that if $\varphi_\mathbf{k}^j(1)$ and $\varphi_\mathbf{k}^j(2)$ are the values of (48) computed for the points (1) and (2) in Fig. 98,

$$\varphi_\mathbf{k}^j(2) = \exp\left(2\pi i \mathbf{k}.\mathbf{R}\right)\varphi_\mathbf{k}^j(1). \tag{49}$$

Here $\varphi_\mathbf{k}^j(1)$ will correspond to the coordinates ρ_1, θ_1 and ϕ_1 of point 1 and $\varphi_\mathbf{k}^j(2)$ to those $\rho_2 = \rho_1$, θ_2, ϕ_2 of point 2. When (48) is introduced in both sides of (49) we shall have one equation on the n unknowns C_{lm}^{kj}, which for simplicity will be called x_i, with coefficients A_i that depend on the radial functions and spherical harmonics. If we now take p pairs of boundary points ($p > n$, say $p = 200$ for $n = 100$) we shall have a system of equations

$$\sum_{i=1}^{n} A_i^r x_i = 0, \quad r = 1, 2, \ldots, p. \tag{50}$$

This means that we take a large number of points on the boundary of the Wigner–Seitz cell and try to satisfy for them the boundary conditions. (In practice, one includes in (50) conditions that establish the continuity of the derivatives at the boundary as well.) Since the A_i^r's depend on the energy parameter ε for which the R_l's are computed, we have one such system of equations for each value of ε chosen. If this were the value that corresponds to the exact eigenvalue E^j in (45) and if the expansion were so large as to yield an exact answer, each equation in (50) would be exactly satisfied. In practice, this will not be the case and for an arbitrary ε we shall have

$$\sum_{i=1}^{n} A_i^r x_i = \delta_r, \quad r = 1, 2, \ldots, p. \tag{51}$$

The problem now reduces to a *least-squares fitting*: ε must be varied until the *residual*

$$S^2 = \sum_{r=1}^{p} \delta_r^2 \tag{52}$$

goes through a minimum. The value of ε corresponding to this minimum is the eigenvalue sought. Standard mathematical techniques exist to solve this minimization problem but attention must be paid to the following. The unknowns x_i satisfy the p approximate equations (51). However, they must also verify an *exact* equation,

$$\sum_{i=1}^{n} x_i^2 = 1, \tag{53}$$

which follows from the normalization condition for the wave function (48). Again, this is a standard mathematical problem, namely that of solving, in the least-squares sense explained above, the equations (50) subject to the constraint (53). As a result, for each system of equations (50) (i.e. for each value of ε) one obtains the residual S^2 which minimizes (52) subject to (53). When ε is varied systematically a curve of the residual versus ε is obtained, as shown in Fig. 99. Since we hold the value of **k** constant throughout this part of the work, the successive minima of this curve correspond to the eigenvalues $E^{(1)}, E^{(2)}, E^{(3)}, \ldots$, for the successive bands at this value of **k**. Of course, all this work must be done automatically in an electronic computer.

The idea of the cellular method was introduced by E. P. Wigner and F. Seitz[†] who used a very simplified form of it. J. C. Slater[‡] showed how

† *Phys. Rev.* **43**, 804 (1933); **46**, 509 (1934).
‡ *Phys. Rev.* **45**, 794 (1934).

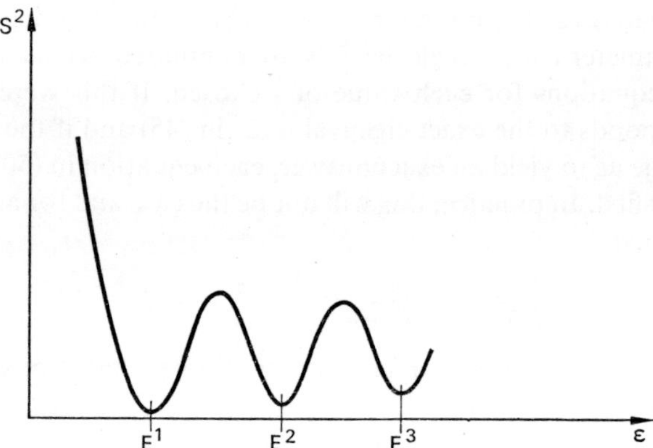

FIG. 99. Fitting of the eigenvalues in the cellular
method.

the boundary conditions could be fitted on the surface of the cell. The
method described in the text is due to S. L. Altmann.†

1. THE POTENTIAL. MUFFIN TINS

In order to solve for the radial wave functions R_l in (47) it is necessary
to have a value of the potential $V(\rho)$. This is the most serious problem in
the cellular method, as in any of the methods which require the radial
functions, such as the augmented plane wave method (see § 5). In practice
one takes $V(\rho)$ to be the potential that corresponds to the metal atom,
which is determined from an exact atomic calculation. However, this is
no more than an approximation since it should be clear from the way in
which we defined $V(\rho)$ that it depends heavily on the metal eigenfunctions
themselves. Various techniques have been proposed to choose the
potential $V(\rho)$ so as to represent as accurately as possible the real
potential in the metal, but uncertainties remain, especially for transition
metals. Some results for some metals are not very sensitive to the
potential field used, but this is far from general and in many cases very
large differences in the band structure appear as a result of comparatively
small variations in the potential field.

† *Proc. Roy. Soc.* (*London*) A, **244**, 141, 153 (1958).

There is another difficulty associated with the potential field. The separation (46) of the wave function in a product of a radial and an angular part is known from quantum mechanics to be valid only when the potential V in the hamiltonian $\mathbf{H}$ of (45) is spherically symmetrical around the centre of each cell. It should be noticed that for this to be the case, it is not enough to take $V(xyz) = V(\rho)$ with the definition of ρ used so far. In fact consider the points 1 and 2 in Fig. 100 which are at the same distance from the centre of the cell on the left of the figure. We have so far defined $V(\rho)$ to be the potential at a given point, ρ being measured from the centre of the cell at which the point is. Therefore at the point 1 the potential is $V(\rho_1)$ whereas at the point 2 it is $V(\rho_2) \neq V(\rho_1)$. This difficulty will arise for all points in the shell of radius between $\rho_{\min}$ and $\rho_{\max}$. Fortunately, the potential is normally a slowly varying function in this region (where the electrons move very much like free electrons) so the error involved is not very large and it is often negligible.

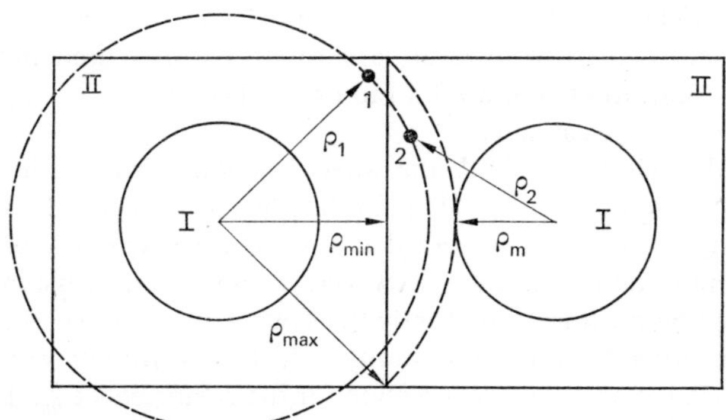

FIG. 100. Cellular and muffin-tin potentials.

In order to eliminate this difficulty the so called *muffin-tin potentials* are often used. They are defined as follows. The Wigner–Seitz cell is divided in two regions. Within the sphere of radius ρ_m (region I) an atomic potential is taken. Outside this region a constant potential is defined. If

$$\rho_m = \rho_{\min} - (\rho_{\max} - \rho_{\min}) = 2\rho_{\min} - \rho_{\max}$$

it follows from Fig. 100 that for all points within the maximum radius ρ_m of the unit cell the potential is spherically symmetrical. For instance $V(\rho_1) = V(\rho_2) = $ constant potential in region II.

It should be noticed that the muffin-tin potential eliminates the mathematical difficulty of the separation of the Schrödinger wave function, at the expense of the use of an *ad hoc* model for the potential. However, this model has some physical foundation, since for many metals one must expect that the electrons move through a substantial part of the crystal space in a uniform potential.

● 5. The augmented plane wave method (APW)

Although in the cellular method either an ordinary, cellular, potential or a muffin-tin potential can be used, the very essence of the APW method depends on the use of a muffin-tin potential. In this sense, it is automatically free from inaccuracies in the separation of the Schrödinger equation into radial and angular parts. This will be the case if the radius of the spherical part of the muffin is taken as $2\rho_{\min} - \rho_{\max}$, as shown in the last section. In practice, however, larger radii are often used and in most practical applications the muffin radius has been taken as $\rho_{\min}$, that is, that of the inscribed sphere.

As in the cellular method, the wave function is written in region I (Fig. 100) as $\sum C_{lm}^{k} R_l(\rho) Y_l^m(\theta\phi)$ (cf. 48), whereas in region II where the potential is constant it is expressed as a plane wave $\exp(2\pi i\mathbf{k.r})$. There is now an important difference, however. These two expressions of the wave function must lead to exactly the same value on the surface of the muffin, in order to satisfy continuity, and this continuity condition is found to lead to an *explicit* expression for the coefficients C_{lm}^{k}. Therefore the wave function is written as

$$\left.\begin{aligned}\chi_{\mathbf{k}} &= \sum_{lm} C_{lm}^{k} R_l(\rho)\, Y_l^m(\theta\phi), \quad \text{in region I,}\\[2mm] \chi_{\mathbf{k}} &= \exp(2\pi i\mathbf{k.r}), \quad \text{in region II,}\end{aligned}\right\} \tag{54}$$

where the C_{lm}^{k} are some determined functions. It should be noticed that although $\chi_{\mathbf{k}}$ is continuous on the surface of the muffin its slope there is discontinuous, since no condition for the continuity of the derivative has been used in (54).

In order to eliminate this discontinuity, we write an improved wave function

$$\psi_{\mathbf{k}} = \sum_{\nu} C_{\nu}\chi_{\mathbf{k}+\mathbf{K}_{\nu}}, \tag{55}$$

where the summation is over all reciprocal lattice vectors $\mathbf{K}_{\nu}$. This expansion is of the same form as that used in the NFE method [cf. (39)].

The coefficients (55) are determined by means of a standard variational calculation, from the solution of the secular equations

$$\det |H_{ij}-ES_{ij}| = 0. \tag{56}$$

The problem is here a little more complicated than usual because the functions $\chi_{\mathbf{k}+\mathbf{K}_{\nu}}$ have discontinuous slope. It can be proved that on account of this, the matrix elements in (56) must be re-defined:

$$H_{ij} = \int \chi_i^*\mathbf{H}\chi_j d\tau + \text{correction term}, \tag{57}$$

$$S_{ij} = \int \chi_i^*\chi_j d\tau \tag{58}$$

[cf. (24) and (25)]. The integrals in (57) and (58) must be carried out over the volume of the unit cell and the correction term in (57) is fully defined in terms of some surface integrals. Once (57) and (58) are computed the determinant in (56) is evaluated as a function of the energy parameter ε (see § 4): the successive values of ε for which the determinant vanishes are the energy eigenvalues for the successive bands corresponding to the value of $\mathbf{k}$ chosen. It should be noticed that whereas the functions χ in (55) have discontinuous slopes, the linear combination $\psi_{\mathbf{k}}$ is smooth, since the variational calculation removes the singularity.

The APW method was introduced by J. C. Slater.[†]

6. Density of states and Fermi surfaces

All the methods described so far in this chapter provide the energy as a function of $\mathbf{k}$ for each band, $E_{\mathbf{k}}^j$. The simplest method to compute the density of states $n(E)$ curve is the following. The range of interest in E

[†] *Phys. Rev.* **51**, 151 (1937).

is divided in small consecutive intervals $\Delta_1, \Delta_2, \ldots, \Delta_n$. A random value of $\mathbf{k}$, $\mathbf{k}_i$ is chosen for which the energies $E^j_{\mathbf{k}_i}$ $(j = 1, 2, \ldots)$ for the successive bands are computed. If the value $E^j_{\mathbf{k}_i}$ falls in the range Δ_m an entry is made under this range. After a certain number ν of random values of $\mathbf{k}$ has been sampled, the entries under the ranges $\Delta_1, \Delta_2, \ldots, \Delta_n$ will be integers $\mu_1, \mu_2, \ldots, \mu_n$, which give the corresponding number of eigenvalues found in those ranges. When ν is very large the ratios $\mu_1 : \mu_2 : \ldots : \mu_n$ become fairly constant and are proportional to the number of states in each range, and by simple multiplication with a constant the values $\mu'_1, \mu'_2, \ldots, \mu'_n$ can be obtained which correspond to $\nu = N$, the total number of values of $\mathbf{k}$ in the Brillouin zone, i.e. to 2 states per unit cell. In this way, a histogram is obtained which gives $n(E)$ as a function of E. From (2.44) the area under this curve up to the value E is the concentration $\mathcal{N}(E)$ of electrons with energies below E. Therefore, by trying out various values of E one can be found for which $\mathcal{N}(E)$ is the known electron concentration in the metal: this value of E is the Fermi energy E_F. (An example of this process will be shown in § 8.)

The Fermi surface is determined as follows. Once $E(k)$ is computed over a suitable grid for $\mathbf{k}$ in the Brillouin zone and E_F is determined, a rough but fairly good idea emerges as to which is the region of $\mathbf{k}$ space for which values of $E(k)$ appear near E_F. Determination of $E(k)$ in more detail in this region gives energy surfaces of constant energy around E_F from which interpolation yields the Fermi surface.

7. Comparison with experiment: heat capacities and de Haas–van Alphen effect

The interest of the calculations described so far is due to the fact that they can be readily compared with experiment: the density of states curve is directly related to the heat capacity of the metal and the Fermi surface to the de Haas–von Alphen effect.

1. DENSITY OF STATES AND HEAT CAPACITY

It follows from the discussion in § 2.11 that the heat capacity of the electrons gas must depend on the density of states at the Fermi surface. In fact, we saw in that section that the heat capacity was proportional to the number of electrons that could be excited into unoccupied levels, which are the electrons at the top of the Fermi distribution. A more detailed argument† shows that the electronic molar heat capacity C_v is given by

$$C_v = \tfrac{1}{3}\pi^2 n(E_F)k_B^2 N_A T, \tag{59}$$

where $n(E_F)$ is the density of states at the Fermi surface, k_B Boltzmann's constant, and N_A Avogadro's number. Equation (59) shows that the electronic heat capacity is a linear function of the temperature. Since other contributions to the total specific heat of a metal are of higher order (see Kittel, *loc. cit.*) it follows that (59) will be the dominant contribution at low temperatures, where in fact the total specific heat is seen to be a linear function of T. From (59), its slope γ is given by

$$\gamma = \tfrac{1}{3}\pi^2 n(E_F)k_B^2 N_A, \tag{60}$$

which is obtained immediately in terms of $n(E_F)$:

$$\gamma = 0\cdot42\, n(E_F)10^{-4}\ \text{cal mole}^{-1}\ \text{deg}^{-2}, \tag{61}$$

where $n(E_F)$ is given as v ryd^{-1} atom^{-1}, v being the number of electron states of either spin.

The agreement between the value of γ calculated with (61) from the density of states at the Fermi surface cannot be expected to be very accurate. Firstly, the calculation of $n(E_F)$ can involve a fairly large uncertainty, specially near peaks. For instance, if the interval Δ with which the histogram of $n(E)$ is computed is, say, $0\cdot02$ ryd, a peak of width $0\cdot01$ could be spread in width by a factor of two and since the area under the $n(E)$ curve should be constant, its ordinate would be reduced by a factor of two. Secondly (60) neglects a number of effects which are now known to be significant. In particular, the interaction between the electrons in the Fermi sea and the lattice vibrations (referred to as *electron–phonon* interactions) can introduce correction factors in (60) which are as large as 2.

† See C. Kittel, *Introduction to Solid State Physics*, 3rd edn., John Wiley, New York, 1966, p. 211. We have introduced a factor of N_A in (59), missing in Kittel, in order to obtain the molar heat capacity.

2. The de Haas–van Alphen effect

de Haas and van Alphen observed the following effect in very pure metals at very low temperatures, under the action of a strong and variable magnetic field H. When the magnetization induced by the field is measured on the metal and the *magnetic susceptibility* χ derived from it, it is found that χ is an oscillatory function of the quantity $1/H$, with a period P. Onsager† proved that P is related to the *extremal cross-section A* of the Fermi surface perpendicular to the magnetic field defined in Fig. 101. Onsager's formula is

$$P = \frac{e}{hcA},\qquad(62)*$$

where e is the electron charge, h Planck's constant and c the velocity of light. [A factor of $4\pi^2$ should be included in the right-hand side of (62) if circular k space is used.]

When units are introduced in (62) it follows that

$$P = 0\cdot 0242 A^{-1} \times 10^{-8} G^{-1},\qquad(63)*$$

where the magnetic field is measured in gauss G and A is the extremal cross-section of the Fermi surface perpendicular to the field and measured in Å^{-2}. If, as in Fig. 101b, there is more than one extremal section perpendicular to the same direction of the field, a beat phenomenon of the two periods will be observed from which each period must be disentangled.

The de Haas–van Alphen effect provides a very powerful technique for studying the geometry of the Fermi surface. It has, nevertheless, its limitations since it provides no information at all about the actual position of the Fermi surface. For instance, if we had a magnetic field perpendicular to the sections of Fig. 74 (p. 166) we would find two periods, one pertaining to the rosette-like section in the first band and another to the cigar-shaped section in the second band. However, nothing would be known about their positions or, for that matter, about their detailed shapes. In order to interpret the experimental results some theoretical model of the Fermi surface must be available. Also, this model

† *Phil. Mag.* **43**, 1006 (1952). See also Kittel, *Introduction to Solid State Physics*, 3rd edn., John Wiley, New York, 1966, p. 294.

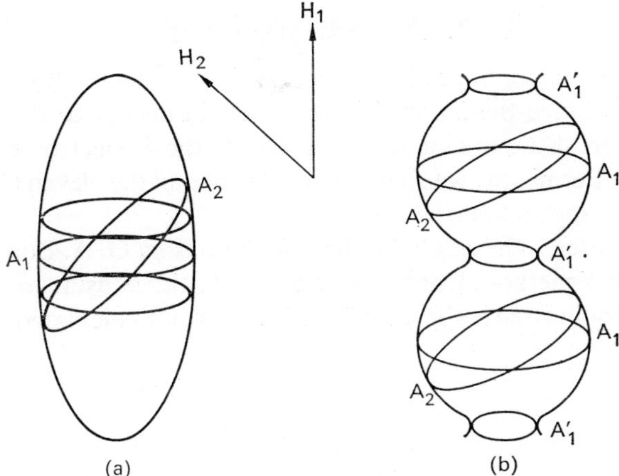

FIG. 101. Extremal cross-sections of the Fermi surface. The cross-sections that are relevant for a given magnetic field are those perpendicular to it. Cross sections with suffices 1 and 2 refer to the fields 1 and 2 respectively. An extremal cross-section is one whose area is a maximum or a minimum with respects to the cross-sections parallel to it. In (a), slices parallel to A_1 are given in order to show that A_1 is a maximum. In (b), A_1' is a minimum and A_1 a maximum.

must be checked by observing many of its sections since it is only then that an overall picture of the surface will emerge. An example of this will be seen in § 8.

PROBLEM. Estimate the areas of the orbits of the Fermi surface shown in Fig. 74 and compute the corresponding periods.

8. The band structure and Fermi surface of copper

As an example of the methods developed in this chapter we shall describe a theoretical study of the band structure and Fermi surface of copper and shall show how the results are interpreted and compared with experiment.

1. THE BAND STRUCTURE

The electronic configuration of copper is $(1s)^2$ $(2s)^2$ $(2p)^6$ $(3s)^2$ $(3p)^6$ $(3d)^{10}$ $(4s)^1$. Since the $3d$ level is very near in energy to the $4s$ one, it would be unrealistic to assume that it is only the $4s$ electron that contributes to the Fermi sea: we shall see in fact that all the eleven $(3d)^{10}$ $(4s)^1$ electrons occupy metallic bands.

We shall discuss the results of the calculations by G. A. Burdick† who used the APW method. Copper is f.c.c. with lattice constant $a = 3\cdot61$ Å, and its Brillouin zone is shown in Fig. 102. Burdick calculated the energy

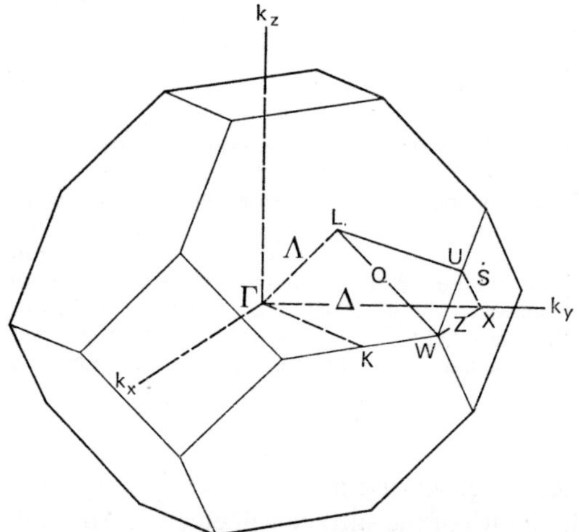

FIG. 102. The Brillouin zone for the f.c.c. lattice.

as a function of **k** along the directions ΓX, XW, WL, $L\Gamma$ and ΓK of Fig. 103a. The symbols that identify the eigenvalues at the points of symmetry and on the lines are a standard conventional notation that permits the identification of the corresponding wave functions. The broken lines along ΓX, ΓL, and ΓK show the free-electron parabolae (see Problem 1). When the wave functions along ΓX are studied it turns out that the parts of the two Δ_1, bands that are near the free-electron parabola are almost

† *Phys. Rev.* **129**, 138 (1963).

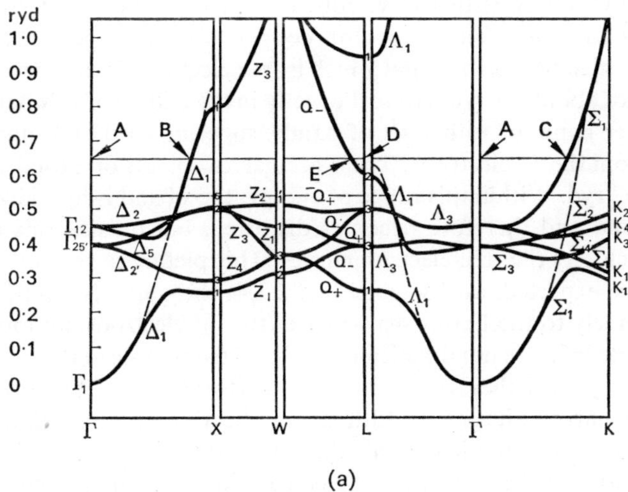

(a)

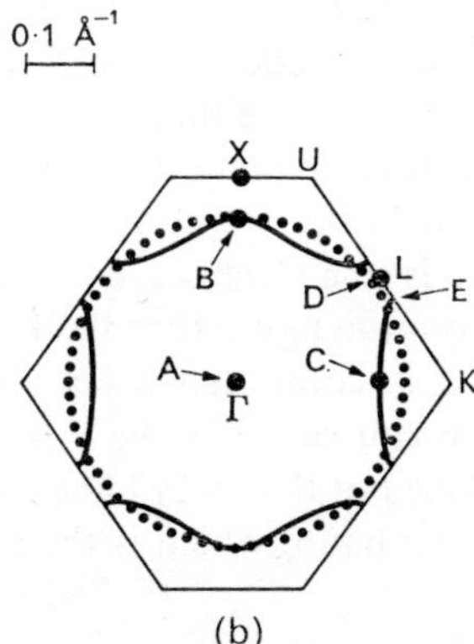

(b)

FIG. 103. (a) The band structure of copper, after Burdick. A constant, -1.039 ryd, must be added to the eigenvalues shown in order to obtain the eigenvalues that correspond to the potential used by Burdick. The full horizontal line is the Fermi energy. The broken lines are the free-electron bands and the broken horizontal line is the free-electron Fermi energy. (b) (110) section of the Fermi surface.

entirely s type, whereas the rest of the bands are essentially d type, slightly hybridized with s wave functions. We can now understand the meaning of Burdick's bands for copper. Broadly speaking, the $4s$ electrons give rise to an s band which is free-electron like and broad, with a width of about 0·8 ryd. Almost exactly in the middle of this band there is another band, or rather set of bands, superimposed on it and which arises from the $3d$ electrons. These are narrow bands of a total width of about 0·2 ryd. Within this range of energy, the s band is hybridized with the d band and therefore becomes distorted, but outside this range it follows closely the free-electron curves. This picture is very much what was to be expected, as discussed in § 2.3.4: the outer $4s$ electrons contribute freely to the Fermi sea whereas the $3d$ electrons, although they participate in it, are not drastically modified from those in the free copper atom. (In fact, if they were entirely unperturbed the d band would be infinitely narrow: its small width is an indication of the small degree of distortion of the atomic orbitals that give origin to the band.)

Some idea of the allocation of electrons in the bands can be obtained by looking at the ΓX direction. The first, Δ_1, band will take 2 electrons; Δ_2, also 2. Δ_5, it follows from some detailed theory, is doubly degenerate and takes 4 electrons. So far 10 electrons, whereas we have 11 to accommodate ($3d^{10}\,4s^1$): the last one goes into the second Δ_1, band which should be half-full, and in fact we see that it is cut through its middle by the Fermi energy so that this band is empty near X. Since the d bands are well below the Fermi surface the whole problem is not very different from the case where there is one electron in a single fairly free-electron band. Thus we can see from the figure that the Fermi energy is not very different from that which is computed for a single free-electron (see Problem 2). In a way, we can say that the d bands do not contribute directly to the Fermi surface but that they have to be taken into account insofar as they perturb the s band, which is the one that emerges at the top of the Fermi sea.

2. THE DENSITY OF STATES

Burdick computed the density of states for copper by a method similar to the one described in § 6. His results are shown in Fig. 104. The Fermi energy E_F is determined so that the area under the $n(E)$ curve up to E_F

corresponds to eleven electrons. The result is 0·66 ryd over the zero of energy used in Fig. 103, and this is the value displayed in this figure. The density of states at the Fermi surface cannot be determined very accurately from the figure, but it is about 3 electrons $\text{atom}^{-1}\ \text{ryd}^{-1}$ which, from (62) leads to $\gamma = 1\cdot2 \times 10^{-4}$ cal $\text{mole}^{-1}\ \text{deg}^{-2}$ in fair agreement with the experimental value 1·7 in the same units.

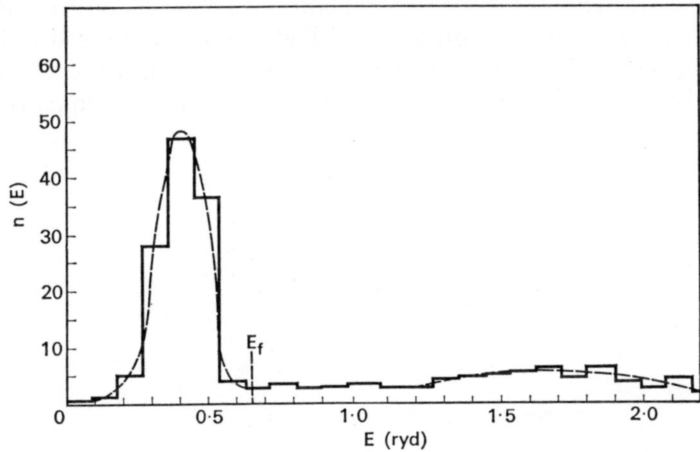

FIG. 104. The density of states for copper, after Burdick.

3. THE FERMI SURFACE AND COMPARISON WITH DE HAAS–VAN ALPHEN RESULTS

A very clear picture of the shape of the Fermi surface can be obtained from the bands in Fig. 103 and the value of the Fermi energy. In Fig. 103b we have drawn the (110) section of the Brillouin zone (see Fig. 102) in exactly the same scale as that used in Fig. 103a. The broken circle is the trace of the sphere whose volume is one half of that of the Brillouin zone. That is, it is the free-electron Fermi surface. (As explained in § 8.1 the copper Fermi surface can be considered to arise from a Fermi sea that contains one electron per atom, the $3d^{10}$ electrons affecting only slightly the quantitative aspects of this picture). From the bands in (a), it can be seen that this sphere must be modified. In the direction $\varGamma X$ the top band is occupied from A to B in (a) and empty from B to X. The radius of the Fermi surface along ΓX in (b) must therefore be the length AB, as shown. Likewise, in the direction ΓK the top band is occupied from A to C in (a) and the length AC has to be taken along ΓK in (b). It can be seen that the free-electron Fermi surface is not distorted at all along ΓX and very little along ΓK. This is understandable because in this direction it is far from the surface of the Brillouin zone. Along ΓL, on the other hand, the free-electron Fermi surface is very near L and in fact in (a) we see that the top band is full at L so that the Fermi surface must be distorted in (b) so as to allow L to be occupied. We ought to have the LK line in (a) but this, unfortunately, was not computed by

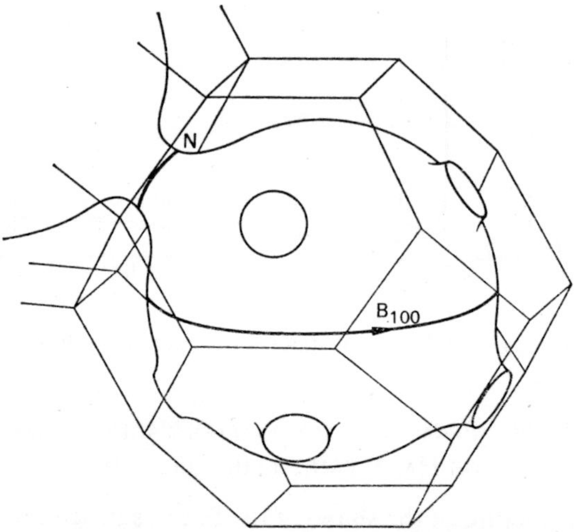

Fɪɢ. 105. The Fermi surface of copper.

Burdick. We can assume that it will not be very different from the LW line (the eingenvalues at W, in fact, are very similar to those at K, in particular the last, vital, one). We see, in fact, that along the LW, or approximately along the LK line, the top band is occupied from D to E so that the Fermi surface must cut the LK line in (b) at E. In fact, as shown in § **4.17.1**, $\partial E/\partial k_n = 0$ along LW so that under our present assumptions $\partial E/\partial k_n$ must be very nearly zero along LK, as depicted in Fig. 103b.

We can now obtain an overall picture of the Fermi surface. It is a spherical object ("belly"), not very different from the free-electron sphere, from which "necks" stick out towards the hexagonal faces of the Brillouin zone. We show this surface in Fig. 105 where the necks as a convenient approximation are represented with circular sections, although they really possess threefold symmetry only.

Two orbits are shown in Fig. 105. B_{100} is a belly orbit that should appear in the de Haas–van Alphen effect when the magnetic field is along [100]. N is the neck orbit that should appear with the field along the [111] direction.

In Table 3 we estimate the areas of these orbits, assumed circular, on obtaining the necessary radii from Fig. 103.

TABLE 3. ORBITS AND DE HAAS–VAN ALPHEN PERIODS
IN COPPER

Orbit	Field	Radius (Å^{-1})	Area (Å^{-2})	P $(10^{-8}\,\text{G}^{-1})$	P exp
B_{100}	[100]	$AC = 0{\cdot}2$	$0{\cdot}13$	$0{\cdot}19$	$0{\cdot}17$
N	[111]	$DE = 0{\cdot}03$	$0{\cdot}03$	8	4
		$0{\cdot}044$	$0{\cdot}05$	4	

The value of P in the fifth column of Table 6.3 is obtained from Onsager's formula (63). In the last column we quote the experimental values found by D. Shoenberg† and the agreement is excellent for the belly orbit and as good as it can be expected for the neck. As a comparison we show in the last line of the table the radius of the neck required to reproduce the experimental period and it can be seen that a very small change in Fig. 103 would produce this result. It should be appreciated that the above is only a rough estimate of the Fermi surface of copper. An accurate theoretical determination of it would require the computation of constant energy surfaces as suggested in § 6.

† *Phil. Trans. Roy. Soc. London*, A, **255**, 85 (1962).

There are various other orbits of the theoretical Fermi surface which ought to be studied. A very interesting one is the "dog-bone" orbit that should appear with the field perpendicular to the [110] direction [i.e. along [110] itself, on account of cubic symmetry], and which is illustrated in Fig. 106. The interest of such an orbit is that, should it be found, it would corroborate the existence of the necks and confirm the interpretation of the [111] period which, in fact, might correspond to a protuberance along the [111] direction rather than a true neck. Unfortunately, Shoenberg did not find the period corresponding to the dog bone orbit.

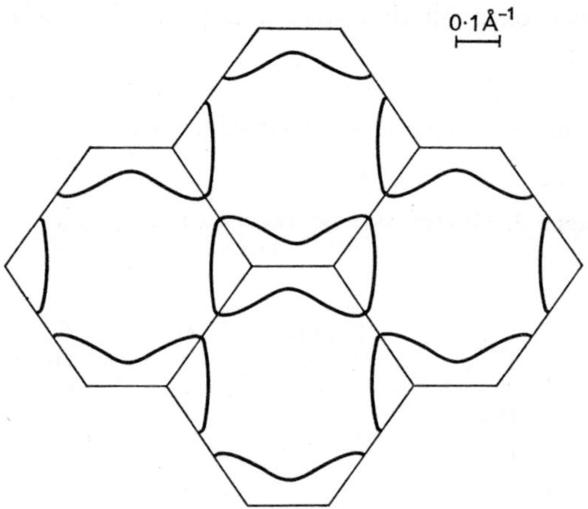

FIG. 106. The dog-bone orbit of copper.
A periodic repetition of the (110) section
of Fig. 103b.

Nevertheless there is no doubt that the necks exist and that the description of the Fermi surface of copper that we have given is essentially correct. First, Shoenberg found the dog-bone orbit in gold, where the Fermi surface is similar to the one described for copper. Secondly, the necks themselves have been found by A. B. Pippard† in copper by an entirely independent method for the study of the Fermi surface (*anomalous skin effect*). Thirdly, the free-electron Fermi surface gets so near L (see Problem 1, § **4.17.1**, and Fig. 103b) that contact with the

† *Phil. Trans. Roy. Soc. London*, A, **250**, 325 (1957).

Brillouin zone surface at L is predicted by a variety of methods, besides the APW calculations of Burdick. In fact, contact of copper's Fermi surface with the Brillouin zone on the hexagonal face was first predicted by a NFE-type calculation by H. Jones.†

PROBLEM 1. The lattice constant of copper (f.c.c.) is $a = 3{\cdot}61$ Å. Find the distance from Γ to X, L, and K and check the corresponding dimensions in Fig. 103. Find the free-electron energies at X, L, and K.

Hint. Use the coordinates given in **(4**.52**)** and the equation **(4**.48**)**.

Answers. $\Gamma X = 0{\cdot}28$ Å^{-1}, $\Gamma L = 0{\cdot}24$ Å^{-1}, $\Gamma K = 0{\cdot}29$ Å^{-1}. The energies at X, L and K are 0·86, 0·64 and 0·96 ryd respectively referred to Γ as zero.

PROBLEM 2. Find the radius of the free-electron Fermi surface for f.c.c. copper ($a = 3{\cdot}61$ Å) and hence its Fermi energy.

Hint. Condition for radius $r: \frac{4}{3}\pi r^3 = \frac{1}{2}\Omega^* = 2a^{-3}$. (From Exercise 2, § **4**.17.1, $\Omega^* = 4a^{-3}$.)

Answer. $r = 0{\cdot}22$ Å^{-1}, $E_F = 0{\cdot}53$ ryd above Γ. Compare with values in Fig. 103.

PROBLEM 3. Estimate the period that corresponds to the dog-bone orbit of Fig. 106.

PROBLEM 4. Find the correct scale in Fig. 103 when circular **k** space is used. Reproduce some of the calculations leading to Table 3 in this space.

† *Proc. Phys. Soc. London*, **49**, 250 (1937).

I

General References

ELEMENTARY

W. HUME-ROTHERY, *Atomic Theory for Students of Metallurgy*, The Institute of
Metals, London, 1960.
J. M. ZIMAN, *Electrons in Metals, A Short Guide to the Fermi Surface*, Taylor &
Francis, London, 1963.
C. KITTEL, *Introduction to Solid State Physics*, 3rd edn., John Wiley, New York,
1966.

MORE ADVANCED

N. F. MOTT and H. JONES, *The Theory of the Properties of Metals and Alloys*, Oxford
University Press, London, 1936.
J. M. ZIMAN, *Principles of the Theory of Solids*, Cambridge University Press,
Cambridge, 1964.
H. JONES, *The Theory of Brillouin Zones and Electronic States in Crystals*, North
Holland Publishing Co., Amsterdam, 1960.

Index

Certain symbols appear after the page number of a reference, the meanings of which are as follows:

A number in parentheses refers to an equation. The letters T, P, E and F (and a number) following a page reference, refer to a table, problem, exercise and figure respectively. The letter n indicates a footnote.

The symbol [3D] before a reference indicates that the subject is treated in three dimensions.

Important symbols used in the text are included with a reference to their defining equation when possible.